Backyard

GUIDE TO THE

Night Sky

Backyard
GUIDE TO THE
Night Sky

Howard Schneider

Foreword by Sandy Wood

NATIONAL GEOGRAPHIC
WASHINGTON, D.C.

Contents

Backyard Guide to the Night Sky

Foreword

THE FIRST TIME I REMEMBER TAKING NOTICE of the night sky was growing up in Texas in the 1960s. I was babysitting my younger brother, who had become cranky from being indoors too long. We went outside, spread a blanket, and lay down and looked up. The Milky Way sprayed its glow all across the dome of the sky. The full Moon was brilliant, which made it not at all difficult to convince my trusting brother that there was, indeed, "a man in the moon."

Years later, when I was about 12 my Daddy woke me very early one morning. "No time to dress," he said, so, still in pajamas, pitch-black outside, we drove to the sea wall in Corpus Christi, Texas. He spread a blanket, poured coffee from a red Thermos, and pointed to a very faint object in the new-Mooned sky. It was so dim that it had to be observed with averted vision (looking near—not at—the subject to see it best). I remember seeing something that seemed to fill a quarter of the sky! My memory says it was a comet, but my research shows that it must have been another object (records indicate there weren't any major comets visible from there that year). What is indisputable is how much I loved being on this nighttime adventure with the man I most adored.

Many years later I took my adorable little mother to McDonald Observatory in Fort Davis, Texas, to view the Leonid meteor shower. Then in her late 70s, but undaunted by the chill and the hard ground, my mother lay close to me atop a thin blanket. For almost two hours, we thrilled to a constant barrage of meteors as they streaked overhead. We clapped, pointed, and oohed and ahhed as the heavens staged one of its spectacular light shows.

The brilliance of stars, planets, and constellations still takes my breath away. And although I cannot deny the thrill of looking through the big telescopes at McDonald Observatory there is another more accessible way: National Geographic's *Backyard Guide to the Night Sky*. It is the perfect tool to get you started on your own lifetime of captivating experiences. Take a clear evening, grab your copy of *Backyard Guide to the Night Sky,* and don't forget the blanket!

Keep looking up!
Sandy Wood
Host of "StarDate,"
a production of The University of Texas McDonald Observatory

Stars leave trails of light as they circle Polaris, the North Star.

How to Use This Book

EACH OF THE TEN chapters found in *Backyard Guide to the Night Sky* is overflowing with information to guide you into the world of astronomy. From the Moon and the Sun, to the stars and the planets, each chapter takes a comprehensive approach to the subject at hand. Not only will you find practical advice on how to find what you're looking for and how to recognize what you're looking at, you'll also find the latest discoveries and scientific explanations to understand the objects and their movements in the night sky.

Alongside the fact-filled narrative, complementary reference elements pepper every page with fascinating facts and figures, the stories behind the stars and planets, short biographies of key sky watchers, and a wealth of cross references and weblinks to direct you to even more information. Look to the Feature essays that cap off chapters for practical tips and advice to help you get more enjoyment and understanding from your astronomical experience.

1 WEBLINK: Websites appear in this location to guide you to resources for more information on a topic.

2 SIDEBARS: There are three different types of sidebar. Sky Watchers tell the stories of key astronomers and scientists. "The Story of" sidebars present legends and lore associated with night sky phenomena. Lastly, "The Science of" sidebars explain the science behind night sky observation.

3 SKYFACTS: These boxes contain fascinating facts, short histories, and cool things to spot in the sky.

4 NARRATIVE: Lively, approachable text thoroughly covers both the practical and scientific aspects of the topic at hand.

5 FAST FACTS: These lists provide a quick reference to the key facts and figures about night sky objects.

6 ICONS: Celestial objects often have a symbol, story, or sign associated with them, which are featured in their specific entries.

7 TABLES: Tables chart key information—such as the layers in the Earth's atmosphere, a constellation's main stars, and annual meteor showers—and if any equipment is needed to view them.

8 STAR MAP: Each of the 58 constellations included in Chapter 8 contains its own star map, created exclusively for this book. To help you locate the constellation and other features, each one highlights the constellation, its main stars, deep sky objects, and neighboring constellations.

9 INTERESTING OBJECTS BOX: In Chapter 8, these boxes appear to give more information on the amazing objects you can find in the constellation—from whirling galaxies to gassy nebulae, from sparkling star clusters to binary stars.

1 WEBLINK

2 SIDEBAR

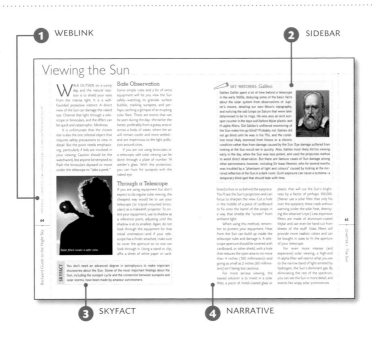

3 SKYFACT

4 NARRATIVE

5 FAST FACTS

6 ICON

7 TABLE

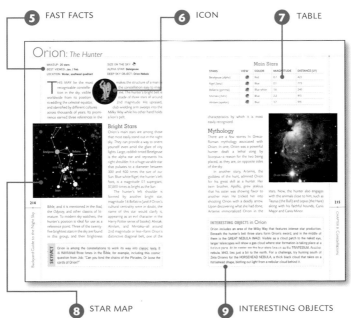

8 STAR MAP

9 INTERESTING OBJECTS

9

Sky Watching Basics

1

• • •

Into the Dark

THINK OF THE NIGHT SKY as one of the earliest sources of entertainment and news. The parade of constellations, the eruption of an aurora, the blaze of a comet—all shaped human culture and gave early civilizations information about issues as critical as when to plant or harvest.

Sky watching remains an accessible hobby, regardless of whether you live in a city bathed in ambient light or in the countryside with dark surroundings. Engaging for a novice, astronomy can become a consuming pursuit, a scientific field where amateurs can still contribute important information.

Attaining expertise, of course, requires both practice and an investment in equipment: high-powered binoculars designed for astronomy; telescopes with computer-tracking, photographic attachments; perhaps even a flip-top backyard observatory. But all that can wait.

SKYFACT

Amateur astronomers have contributed much to the field, and Leslie C. Peltier was among the most prolific. Working in the first half of the 20th century, he discovered ten comets, and made some 132,000 observations that tracked the shifting brightness of variable stars.

Venturing outside is the first step

Getting Started

Beginning requires nothing more than a walk outside and an open sky. The naked eye remains the most important piece of "gear," and a focus on easily seen objects is the best way to build a foundation. Your first tasks may appear simple but will reveal much: Study the phases of the Moon. Track the Sun's changing path over a few months. Learn the behavior of the brightest objects visible. These projects will demonstrate how the heavens work and allow you to build a better appreciation of the fainter and more distant things you'll see later.

Where to begin? Darker is obviously better—the view from a country field or suburban park will show more than a city street bathed in glare. If stargazing at home, turn off as many lights as possible. If heading into the country, find an observation point that keeps the city lights to your back (in other words, if you want to focus on the southern sky, go south of town). Avoid unshielded streetlights or security lamps. A black cloth or jacket held over the head can help block some of the invading light.

Be Prepared

Give your eyes time to adapt to the dark—and then keep them adapted. This process takes a minimum of 15 to 20 minutes and can be undone with a brief glance into a car headlight.

Regular flashlights also cause problems. Since you'll need light to read your guidebook (like the sky charts in Chapter Seven), tape a piece

THE SCIENCE OF Night Vision

Before sky watching, 20th-century astronomer Clyde Tombaugh would prepare his eyes by sitting in the dark for an hour. It can take that long for the eye to reach its peak night-vision ability. In the dark, the pupil will widen and open to let in more available light. Low- or no-light conditions allow photochemicals in the rod and cone cells to regenerate to their maximum level—a process that can take up to 45 minutes for the rods, the receptors most responsible for night vision. If you don't have an hour like Tombaugh—who would later go on to discover Pluto—even 15 or 20 minutes in the dark will help.

of red cellophane over a regular flashlight lens, or buy a red-lensed light. Red light, especially dim red light, hinders night vision much less than does white light.

Next, watch the weather strategically. For the novice, clear air is the best. Take advantage of nights when a cold front or late afternoon storm has cleared out humidity and haze. Also, look for developing high-pressure zones; they create clear skies.

Finally, serious sky watching takes patience, and staying comfortable is key. Plan to keep fed and hydrated. And because even seemingly warm summer nights can turn chilly after hours of stargazing, pack layers of clothing appropriate to the season. Shoes may end up soggy as dew condenses; waterproof gear can be expensive, so spread a blanket on the ground to prevent the problem. And to relax at the best viewing angle, bring a lawn chair.

13

Getting Oriented

EARTHLY reference points are familiar—the marks on a compass; the Equator dividing the planet into hemispheres; the North and South Poles; measurements of longitude and latitude to fix location. Astronomical conventions for locating objects in the sky use similar terms. Applying them requires both a conceptual leap and an appreciation of the fact that what you see and where it appears depends on where you are and when you are looking.

The Celestial Sphere

It is now known that Earth is not at the center of the cosmos, but star maps are still oriented as it it were. It will help, in fact, to imagine that Earth is enveloped by a giant globe called the celestial sphere. Its equator—the celestial equator—is parallel to Earth's. Lines extending out from Earth's North and South Poles pass through the sphere's north and south celestial poles.

Next, visualize that all the stars are attached to this sphere. Stars can be found using two coordinates called declination and right ascension. The first is akin to latitude—how far the star is above or below the celestial equator, measured from zero (at the equator) to plus (for the north) or minus (for the south) 90°. Right ascension is akin to longitude. It is divided, just like

Earth's day, into 24 hours, with each hour equaling 15° of circumference. The starting "zero hour" is a north-south line that marks where the Sun hits the Earth's Equator on the vernal equinox. Now, set the celestial sphere revolving from east to west. While the Earth is actually spinning from west to east, the effect is the same—stars pass overheard from east to west.

It is critical to know both your approximate latitude and the reference latitude of any star guide you're using.

14

Backyard Guide to the Night Sky

Star trails circle the north celestial pole.

(Charts in this book, for example, are based on a view from latitudes close to 40° north.) If you were standing at the North Pole, you could not see the stars located south of the celestial equator. Travel south, however, and more of the celestial sphere below the celestial equator comes into view.

Finding North

In the Northern Hemisphere, the star Polaris is the starting point. Located almost directly above Earth's North Pole, it is visible all year. Find it by spotting the Big Dipper and extending an imaginary line through the two stars at the far edge of its "bowl." Polaris is approximately one "dipper's length" away. Face it, and you face north—south is behind you, west on the left, and east on the right.

Polaris's height in the sky also indicates latitude. At the North Pole (90° N), the star is overhead—at a 90-degree angle to the horizon. Moving south, the star drops lower in the sky. From the northern U.S. and southern Canada, corresponding to latitudes of around 40° to 50° N, Polaris is roughly halfway up the northern sky.

🔭 **SKY WATCHERS:** Ptolemy

Like many early thinkers, second-century Greek astronomer Claudius Ptolemy was deceived by the apparent motion of the Moon, Sun, and stars into thinking that the Earth was at the center of that movement. It was possible, in fact, to construct an Earth-centered model that accounted for motion quite accurately. His system of epicycles—smaller and larger orbital circles placed around other circles—accomplished part of the task. He also had to assume that Earth was slightly tilted on its axis, something we now know to be true even in our Sun-centered reality.

The Ecliptic

YOUR VIEW of the sky depends on more than just location. What we see also depends on the time of year, and changes slightly on a daily basis. Remember that one of the objects on the celestial sphere is the Sun, which appears to follow its own course as that imaginary globe revolves around us. The apparent path of the Sun through the year is called the ecliptic—a line that reflects Earth's orbit around the closest star. The ecliptic determines much of what is visible to an earthbound observer.

Changing Views

With a few bright exceptions at sunrise and sunset, the celestial objects overhead during daylight hours (and remember, the stars are always out there) are invisible to the unaided eye because of the Sun's brightness. But as Earth orbits around the Sun, our view constantly changes. Each nightfall brings a slightly different patch of the "sphere." Stars invisible in June, when they are washed out by the noonday Sun, will be riding in the night sky six months later when Earth has moved to the other side of its orbit. Star charts are typically monthly or seasonal to take this changing view into account.

The ecliptic helps explain Earth's seasonal weather, the changing amount of time the Sun spends in the sky each day, and the differing heights

SKYFACT

The apparent motion of the Sun and stars provided the earliest ways to tell time, with the westward journey of the Sun giving a sense of the day's progress, and (at least in the Northern Hemisphere) the rotation of the Little Dipper around the North Star serving the same role at night.

On a "white night," the Sun dips low but does not set.

The Sun, Venus, the Moon, and Jupiter in a multiple-exposure photo

it reaches between rising and setting. Earth is tilted about 23.5 degrees off of its plane of travel around the Sun, but the "direction" of that tilt remains constant. At one spot in the orbit, the Northern Hemisphere tilts toward the Sun and enjoys a season of longer days as the Sun arcs high into the sky. Six months later the north tilts away from the Sun, and the geometry of daily orbit brings the Sun into view for less time and at a lower angle to the horizon. The situation, of course, is reversed in the south. In the spring and fall, when the direction of tilt is "flat" to the Sun, the amount of sunlight is more evenly distributed between the hemispheres.

Ecliptic Objects

The ecliptic also defines one of the most interesting bands of the sky. The solar system evolved from a cloud of spinning gas that created the Sun. In the process, heavier material settled out into what would become the planets, all orbiting in roughly the same plane and visible along it at different times. The constellations of the zodiac follow roughly the same path, with Earth's journey around the Sun bringing a new one into view every month.

THE SCIENCE OF White Nights

St. Petersburg, Russia, may have a dreary image for its cold climate, but from early June to early July the city celebrates the near-constant daylight—the "white nights"—that sets in around the summer solstice. Because of its location at nearly 60° north latitude, St. Petersburg is pointed sharply toward the Sun during the summer. The Sun, as a consequence, never drops more than a couple of degrees below the horizon, creating a sense of twilight until it rises the next day. At the North Pole the Sun is in the sky from the time it rises on the first day of spring until it sets on the last day of summer.

Star Hopping

COMPUTERS have taken some of the craft out of locating objects in the sky, with software and motor drives available on even amateur telescopes to find thousands of objects contained in pre-installed databases. But for those not ready to make that kind of investment, or those who wish to bring a do-it-yourself spirit to the hobby, the rudiments of star hopping will help guide you through the night sky.

Finding Your Way

Whether using a telescope or the naked eye, star hopping provides a sort of road map that uses brighter objects to point the way to fainter ones. The Andromeda group of stars, for example, provides stepping stones to help spot the Andromeda galaxy, one of the closest neighbors to the Milky Way and the deepest object visible (when conditions are right) to the unaided eye. In the late northern summer sky, bright Alpha Andromedae forms one corner of the asterism, the Square of Pegasus. It also marks the "tip" of the head of the Chained Lady—Andromeda herself, in constellation form. Gamma Andromedae and Beta Andromedae follow in line. From there, move up to the next reasonably bright star, and then to the one after that, and you'll be in the neighborhood of the galaxy, which looks like an oval, hazy patch about as wide as a fingertip. It will take a moonless night and dark conditions to see it without equipment, however.

Vega, high in the northern sky and the second brightest nighttime star in the hemisphere, helps point out Deneb and Altair, partners in the summer triangle. The bowl end of the Big Dipper points to Polaris, the North Star, but the

SKYFACT

Between the tug of the Sun and the Moon, Earth wobbles slightly as it rotates on its axis. The motion is called precession.

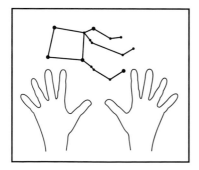

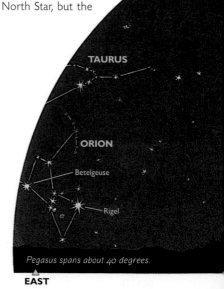

TAURUS

ORION

Betelgeuse

Rigel

Pegasus spans about 40 degrees.

EAST

dipper's handle is helpful too: It is the start of the "arc to Arcturus," the brightest northern star and the center of the constellation Boötes. From Arcturus, in turn, you can "speed on to Spica," in the constellation Virgo.

Degrees of Distance

You don't need an expensive gadget to calibrate distance. Typically measured in degrees, distance can be estimated with your own hands (bottom left). Outstretch your arms and hold out your hands against the sky. At arm's length, a thumb covers approximately 2 degrees of sky, a fist about 10 degrees and a full out-

As they settled across a massive geographic area beginning perhaps 3,000 years ago, Polynesian sailors relied on an encyclopedic mental map of local islands and stars. The horizon was divided into sections based on the rising and setting of bright stars like Antares and Vega. Knowing that a certain island lay on a line between the rising point of one star and the setting point of another allowed local navigation. Their system became so extensive that they seem to have made it as far east as Chile.

stretched hand about 20 degrees. These approximations can be used to estimate the width and height of constellations. Pegasus, for example, covers about 40 degrees, about two hands, on the sky.

To calibrate your own hand measurements, take a look at the Big Dipper. From end to end, it covers about 25 degrees. The two "pointer stars" in the bowl end are about 5 degrees apart, and it is roughly 28 degrees from there to Polaris.

ANDROMEDA
ARIES
PEGASUS
PISCES
AQUARIUS
Altair
CAPRICORNUS
AQUILA

SOUTH   WEST

Keeping Time

GREAT NEWS! There's a lunar eclipse on a nice summer Saturday night—plenty of time to watch it develop without worrying about work the next day. You set up the telescope and wait. And wait. No eclipse? Oh, it happened, but was visible from Europe and Africa, which experience night many hours before the Sun sets in North America.

Getting in Sync

Time—along with date and latitude—is important to track when looking for celestial events. Given the world's two dozen time zones, it is important to

Time zones differ worldwide.

stay in sync with the guidebooks and charts. The issue is not so critical for the star positions: If a given constellation will cross its highest point at midnight, that's true for all time zones—it simply happens three hours later in Seattle than it does in New York. But for discrete events like eclipses, which start and end at specific times, synchronization is vital.

One standard is to use Universal time (UT), the local mean time in Greenwich, England. When the move toward a system of world time zones began in the late 1800s, Greenwich was chosen as the prime meridian—set as 0° longitude and the starting point to divide the globe into segments where time would advance one hour (moving east) or fall back one hour (moving west) every 15 degrees. Astronomers use UT to identify moments when celestial events occur.

If you see it referenced in a guidebook, you'll need to convert it to your local time. In North America, subtract five hours for eastern standard time, six hours for central standard time, seven hours for mountain standard time, and eight hours for Pacific standard time. When daylight savings time is in effect, in each case subtract one hour. Occasionally, converting changes the date to the previous day: for example, an event at 03:20 UT on November 25 occurs at 10:20 p.m. November 24 in eastern standard time.

Days on Earth

Time itself is an astronomical construct. Technically, a "day" on Earth

---

SKYFACT

Today, coordinated universal time relies on atomic clocks that keep time based on the vibrating rhythm of cesium atoms. To account for the gradual slowing of Earth's spin (caused by the Moon), a leap second is periodically added to keep them tuned to within 0.9 second of Universal time.

For the ancient Egyptians, astronomy was time. For daily use, they developed a sundial that used a straight base, notched to mark increments of time, and a raised crosspiece to cast a shadow. But they were adept at marking longer periods as well. They were the first to set a yearly solar calendar at 365 days and had marked the movements of 43 constellations and five planets as early as the 13th century B.C. Events like the annual rising of the star Sirius helped anticipate the annual flooding of the Nile and were used to coordinate agriculture.

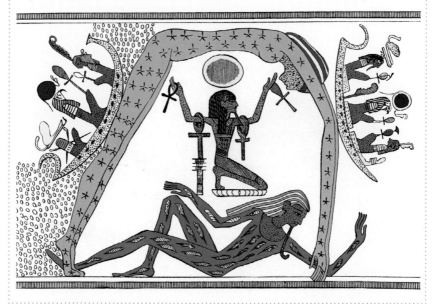

is the time it takes for the planet to turn once on its axis. That is measured in reference to the Sun's passage from one meridian on one day, and its return to the same meridian on the next. Because the planet's orbital speed varies over the course of the year, the 24-hour clock is actually an average of these "solar days," and is referred to as mean solar time.

That differs from the time it takes for the Earth to rotate once relative to the distant stars. A sidereal day, or star day, is about 23 hours and 56 minutes.

Measured on a standard Earth wristwatch, in other words, a star that passes a given meridian at 9 p.m. one night will return there at about 8:56 p.m. on the next. That happens because our daily change in position relative to the Sun, though slight, is far greater than it is in reference to stars that are millions of light-years away. Because the Earth is moving along its ecliptic path as it is turning, the planet must turn a bit farther than once around—about four minutes' worth—for the Sun to pass the same reference point.

Light Pollution

U NDER GOOD conditions from a rural location in North America, there may be perhaps as many 1,500 stars visible to the naked eye. Overhead photos of the Earth at night, however, show that the "best of conditions" can be hard to find, particularly for the people who live in urban or suburban areas.

Atmospheric effects, light pollution, and other constraints can reduce the number of visible stars—perhaps considerably. The viewing limit of the eye may be cut by a factor of around 40, from a 6 on the stellar magnitude scale down to 2, possibly even washing out guidepost Polaris.

Saving Starlight

Light pollution's spread affects the night view even far away from the city limits as more lamps appear along highways and fixtures burn through the night in parking lots and shopping

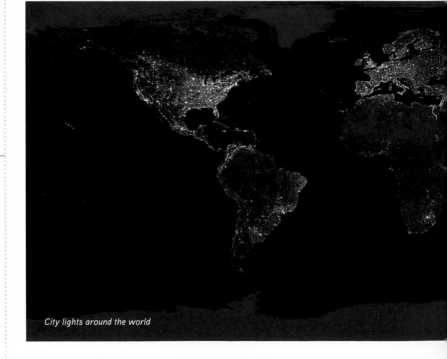

City lights around the world

centers. Much of that light is wasted, cast up into the sky or spread far away from the very area it is supposed to

illuminate. Shielded lights would serve the same purpose at a lower wattage, and several lobbying groups have begun to argue in favor of more efficient lighting ordinances.

Protecting the night sky has prompted creation of dark-sky preserves at places like Cherry Springs State Park in Pennsylvania. The surrounding government has adopted ordinances to limit light pollution, and the park bans the use of white lights if people are observing the sky.

Finding a Dark Sky

Overcoming light pollution in your neighborhood may take some scouting. Ideally, you'll want to keep the city lights to the rear. If you want to study the southern sky, then drive south from the city. Go far enough away so that the city's glow doesn't creep higher than ten degrees or so into the sky.

Avoid spots where specific sources of light—like car headlights or street lamps—prevent the eye from fully adapting to the dark. Portable shields can be constructed from plastic pipe and black cloth if you find a spot that has one or two point sources of light.

If using a telescope, filters are available to block wavelengths common in streetlights. But this equipment is advanced: Remember that it's best to scout your viewing location before making an investment. Consider the ease of assembly and how close you'll be able to park to your observing site: Multiple trips to the car and a long setup time are a strong disincentive even if the viewing spot is ideal.

Sky Watching in the City

New York City at night

24

B ETWEEN the buildings blocking the horizon and the light pollution, urban sky watching might seem a lost cause. While the subtle objects of deep space or low-to-the-horizon comets and stars might be out of reach, there is plenty left to study—particularly for the novice.

Bright Objects

The Sun and Moon, for example, are the two most obvious objects in the sky, and urban residents are at no special disadvantage when studying their movements. Learning the basics of these two bodies will teach you the basics of the solar system.

A warning: Special procedures and equipment, described in Chapter Three, are needed to watch the Sun safely. The intense light from that closest star, viewed through optical equipment or directly through the unaided eye, can quickly ruin your vision. But with the right precautions, Sun gazing can be a hobby in itself. The Moon, as well—big, close, and well lit—will cut through urban glare and offers a rich amount to study: its surface geography, its phases, and its eclipses.

Beyond that, Venus reaches magnitudes as bright as -4.2, while Jupiter and Mars shine as brightly as -2.9 and -2.8, respectively—well within the limits of city viewing. Mercury can also be bright, but its orbit close to the Sun limits the times when it can be spotted. Saturn, with a magnitude of roughly 0.7, can also be seen unaided even in bright conditions.

There are, in addition, 16 stars with a magnitude of 1 or less. Think of them as building blocks. From an observing site in the city, they will stand out in their solitude, and be easier to use as star hopping guides when you begin to acquire equipment or move to darker locations. Even in the city, the use of binoculars or a telescope can bring thousands of stars into view—if you set yourself up correctly.

Light Pollution

There are ways, for example, to battle light pollution. First, look at what is under your control. When observing from your own yard, turn off porch and interior lights, and ask the neighbors to do the same, if you feel comfortable. If necessary, use the shade of a fence or bush to shield what light you can—assuming, of course, that it does not impede the view you are trying capture. An extension tube at the end of the telescope can make an extra-long lens hood. Local parks may offer a bit of refuge. If you find a spot with a stray streetlight, a portable curtain can be built with lightweight plastic pipe and black cloth.

Finding objects in the absence of visible reference stars will be hard, an argument for the computer-guided telescopes with coordinates preloaded into the software. Urban viewing may

also require a different choice of eyepiece for the telescope. The diameter of the ray of light leaving the eyepiece is called the exit pupil. A smaller exit pupil and higher magnification will help counter light pollution by darkening the background and increasing the contrast with the stars or other objects you are trying to see.

And be strategic: Urban horizons are cluttered, and there is more light there. Hunt for objects that are as high overhead as possible, positioned away from the glow of the city.

SKY WATCHERS: Star Parties

Urbanization may be wiping out the night sky, but it has also helped expand the resources available to see it and learn about it. Local museums are likely to sponsor star watching events, or star parties, where novices and experienced amateur astronomers view the skies. A call to the astronomy department of a university or college can unlock a wide variety of resources—from a nearby planetarium to a research telescope that has hours for public viewing. Major cities are likely to have local astronomy clubs as well. The website for *Astronomy* magazine, under its community section, has a list of local clubs, while the ClearDarkSky website (*www.cleardarksky.com*) offers local watching conditions at popular dark-sky sites around the country. Attending one of these events is a good way to learn about equipment and viewing techniques.

Binoculars

THE SIDEKICK of bird-watchers and sports fans, binoculars have little of the history or romance associated with telescopes. But consider: Telescopes limit you to one eye, an awkward situation for the brain. The amount of sky seen through the lens—called the field of view—is narrow, making it harder to find the target. Telescopes can be bulky and hard to set up—a disincentive if the night is chilly or time is short. Good-quality equipment can also be expensive.

For ease of use, portability, and economy—all while still adding remarkably to the range of what you can see—a standard pair of binoculars is a recommended first piece of equipment for the beginning astronomer.

What to Shop For

There are some important points to keep in mind before purchasing. For example, weight matters. You'll be holding the glasses and looking up, so some of the heavier models may become wearing while scanning the sky. You can, of course, invest in a tripod, but the ability to take a casual scan overhead without a lot of setup time is one of the advantages of binoculars in the first place.

The next step is choosing the right size, as measured by the level of magnification and the size of the objective lens (the outer lens). You'll likely see both stamped on the glasses somewhere—something like 7x35 or 10x50, with the first number being the magnifying power and the second the diameter of the lens in millimeters.

For general viewing a 7x50 configuration is a recommended choice. The 50-millimeter-diameter objective lens is a sort of consensus size for basic astronomy, large enough to gather the

SKY WATCHERS: Best Binocular Sights

Though hardly exotic, binoculars bring a wealth of objects into view. You can see the larger craters on the Moon. Jupiter's four biggest moons can be tracked in their orbits. Mercury, elusive to the naked eye, can be spotted more easily, and the more distant planets Uranus and Neptune can also be detected. The wider field of view of binoculars will provide a more dramatic view of comets, compared with the narrow range of a telescope. Numerous deep sky objects are accessible as well, and for some of them—large star clusters like the Hyades and Pleiades in Taurus, for example—the wide field of view is a benefit. Star fields along the Milky Way make glorious sights in binoculars, and it's fun to spot several other galaxies, including Andromeda and the Pinwheel galaxy.

Moon with naked eye

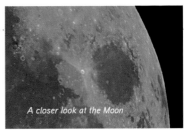
A closer look at the Moon

light needed for serious viewing but compact enough to remain on a neck strap. The field of view is wide, and the image produced by a 7x magnification can be steadied by hand. Remember that binoculars also magnify the effects of trembling fingers: Even the jump to a 10x50 pair might require a tripod to control the jiggling.

The 7x50 also has the advantage of providing an exit pupil—the beam of light coming from the eyepiece—that roughly matches the size of a fully dilated human eye, about 7 millimeters (0.275 inch). The exit pupil equals the size of the objective lens divided by the level of magnification. Although people lose some dilation as they age—meaning a bit of that 7 millimeters of light will be wasted—the larger exit pupil also makes it easier to keep the eye aligned.

Other Considerations

The binoculars' prisms should be made of BaK-4 glass, as opposed to BK-7. It will cost more but is worth it for the extra brightness. You can tell by holding the glasses at roughly arm's length and examining the eyepieces. If the bright circle of light you see—the exit pupil—has any gray edges, the binoculars you are holding use BK-7 glass.

Shop also for glasses with multiple-lens coatings. They transmit light better (you can tell by shining a flashlight through the outer lens and tilting the glasses back and forth. Coated glass will look blue or green). Last, while in the store, test for alignment, also called collimation: Focus on an object, then alternate closing your eyes; the image should stay stable. Hints of rainbow color around bright reflections or double images indicate lower quality optics.

You may also encounter giant astronomy binoculars with lenses of 80 or 100 millimeters (3 or 4 inches)—or more. Their price can equal or exceed that of a telescope. They are also far too heavy and powerful to steady by hand, and will require a sturdy tripod. For novices, this is not so much a step toward a telescope as a substitute for one. Buying a higher quality pair of smaller glasses might be better option for those first tours of the sky.

SKYFACT

Think binoculars aren't powerful enough to reveal much about the objects in the night sky? Then consider this: A good pair of modern binoculars are stronger than the telescopes used by 17th-century astronomer Galileo when he discovered Jupiter's moons and the phases of Venus.

Sky Watching With Kids

T HE SKY is alive with myths and monsters and gods, whether it's the legends behind the zodiac, the names of the planets, or the wide variety of tales associated with phenomena like the aurora borealis. Science fiction, meanwhile, brings us the promise of a visit by a friendly extraterrestrial, or a menacing attack launched à la H. G. Wells.

For kids, at least at first, star watching may be less about the science, and more about the mystery of our place in the cosmos. To a young mind, the broad brush of a dark sky may be more intuitively satisfying than the hunt for a deep space object.

That sense of imagination can make for some wonderful family outings but may involve compromise regarding the type of equipment being used and the organization of time spent in the field. Most children will not go on to be astronomers or even sustain astronomy as a hobby. But an appreciation of the basics will stay with them for a lifetime, and for that no fancy equipment is needed.

A Kid's-Eye View

Binoculars, for example, are a more accommodating way to look at the sky when children are involved. They require no setup and, being less expensive, will cause fewer worries about accidents than a telescope. If you have a choice, opt for a smaller and lighter size better suited for kids' hands. With a wider field of view, binoculars bring in more sky and may better suit the things a child wants to view and the

way he or she wants to look at them. Have at least a pair (or more) on hand to share.

Picking the Right Time

The timing of trips with kids can be tricky. The late setting Sun of summer may leave little time before they begin to tire: Go before sunset, and plan the outing around appropriate objects—a first quarter Moon, perhaps, or Jupiter or Saturn when it is near opposition. When the timing's right, see who can spot Venus first.

Winter offers an earlier night and new star patterns to view. But kids are more susceptible to cold than adults, and will likely need an extra layer or two of clothing (and a Thermos of hot chocolate) to be comfortable.

For a child who has become more serious about the hobby and is ready for a telescope, consider some of the easier-to-use models, particularly those with Dobsonian or computer-driven mounts (described on p. 261). Ideally, you'll want to invest in something that the child can independently set up and use.

Last—and most important for any viewing when the Sun is still in the sky—impress on children the warnings about how the Sun can ruin eyesight, and watch your equipment at all times to be sure no one hazards a look.

Sky Watching Projects for the Family

OBJECTS	TIMING	WHAT TO LOOK FOR
Man-made satellites	Year-round, just after sunset or just before sunrise	Starlike lights that quicky and steadily move across the sky (see p. 87)
Venus	Check almanacs and guides to see if Venus is best seen in the morning or evening.	A steady bright orb. Does not twinkle like a star (see p. 102)
Big Dipper	Year-round	The most easily recognizable star pattern (see p. 46)
Sirius	January, February	Brightest star in the sky. Belongs to the constellation Canis Major (see p. 182)
Perseids	July-August	Bright "shooting stars" originating near the constellation Perseus (see p. 255)

Families can enjoy stargazing year-round.

Naming the Sky

Northern constellations from a 1708 star atlas

DIFFERENT CULTURES, different perspectives, different names: Polynesian seafarers created their own labels and legends for the stars, as did Native Americans, and the Greek and Arab astronomers who gave us many of the names in common use today. What awaits any amateur is a hodgepodge of titles—with hundreds handed down over history and retained, thousands compiled in catalogs as telescopes grew more powerful, and millions assigned by scientific convention in the space age.

SKYFACT

Some commercial companies have offered star names for sale as gifts. The International Astronomical Union has disavowed those offers, noting on its website that "like true love and many other of the best things in human life, the beauty of the night sky is not for sale."

THE STORY OF Shakespeare's Moons

Uranus is named after a member of the Greco-Roman pantheon, but the stargazing Herschel family (patriarch William discovered the planet in 1781) pushed the naming of its moons in a new direction. For the first four named moons, son John Herschel used the names of fairies and sprites. The initial ones, Titania and Oberon from William Shakespeare's *A Midsummer Night's Dream*, and Ariel and Umbriel from Alexander Pope's "The Rape of the Lock," were followed by the eventual discovery of 23 others, the names pulled mostly from Shakespeare.

Though the International Astronomical Union has established guidelines for several different compilations of star names (which, for order's sake, have become strings of letters and numbers, as opposed to elegant names like Vega), amateur observers at first need only be familiar with the proper names and a few more modern naming conventions.

Stars & Comets

Many different civilizations have recognized star patterns and given them names and mythology. In the West, the 12 major constellations of the zodiac and three dozen others were handed down over the centuries, indexed most notably in Ptolemy's *Almagest*. Others are of more recent invention, charted as the Southern Hemisphere was more fully explored.

Several hundred of the brighter stars also acquired proper names over the centuries. Arab astronomers provided much of that lexicon—from Acamar to Zuben Eschamali. Stars sometimes carry other names on charts that use their relative brightness in a constellation: Betelgeuse ("shoulder of the giant" in Arabic) is also Alpha Orionis (the brightest star in Orion).

Comets traditionally have been named for their discoverers (comet Halley being the most famous), though a modern designation system also exists to give the approximate date of discovery and some of the key qualities. For some deep space objects, more poetic or descriptive names still attach—the Crab Nebula for its color and shape, for example.

Checklists & Catalogs

Two catalogs provide an inventory of deep space objects to look for. French astronomer and comet hunter Charles Messier, attempting to list things that were not comets, created an index of galaxies, nebulae, and other items that hobbyists use today to guide their deep sky journey. They are given M, or Messier, numbers. With modern tinkering the list stands at 110.

An overlapping and much larger list is the New General Catalog, put together more than a century ago by Danish astronomer J. L. E. Dreyer and including (with subsequent supplements) more than 13,000 items. The NGC-numbered items include Messier's list, but many of the others will require a larger telescope to view.

How High the Moon?

IS THE UNIVERSE infinite? Can it be measured? Knowledge about the scale of the universe now extends out to billions of light-years—with the Hubble Space Telescope and other devices trying to peer to the very beginning of time. Indeed once incomprehensible interstellar distances have become more and more precise—though the scale and reference points remain exotic.

Units

The total distance between the Earth and the Sun is 93 million miles (149.7 million kilometers). The entire distance from the Earth to the center of the Sun forms one of the basic scales of measurement—the astronomical unit, or AU. Neptune, the outermost planet, orbits at an average of about 30 AU from the Sun. Going farther to the very edge of our solar system, we find the Oort cloud of galactic debris, about 100,000 AU from the Sun's center.

An astrolabe

The units only increase. A light-year is the distance light travels in a year—and at roughly 186,000 miles (299,338 kilometers) a second, that amounts to about 5.5 trillion miles (8.9 trillion kilometers). Excluding the Sun, the closest star to Earth is Alpha Centauri, about 4.3 light-years away, while the Milky Way galaxy spans about 150,000 light-years.

Methods

How are such distances calculated? Direct measurement is impossible, but a number of methods can infer distance, and often different methods are used to corroborate the results produced.

One basic method for measuring the distance involves the optical illusion known as parallax—the apparent motion of an object caused by a shift in the observer's perspective. As Earth orbits the Sun, stars seem to shift position. For closer stars, the shift is large enough to estimate distance. In observations taken six months apart, scientists use a ratio between the distance the Earth traveled (two AUs, from one side of Earth's orbit to the other) and the angle of the star's apparent shift. The distance is usually measured in parsecs—the distance that yields a parallax of one second of arc, or about 3.26 light-years.

Another method involves using a spectroscope to analyze the light emitted from a star to determine its intrinsic brightness. Comparing this with the way it actually appears lets astronomers calculate the star's distance from Earth. Similar techniques involve using the cyclical brightening and dimming of variable stars. For the most remote galaxies, astronomers use one kind of exploding star, a Type Ia supernova, as the "standard candle" for estimating the galaxy's distance.

THE SCIENCE OF Hubble's Constant

Edwin Hubble, whose name is borne by the famous space telescope, was the first to detect the individual stars of the Andromeda galaxy. His work also produced a basic tool used for estimating distances to galaxies. Using the phenomenon known as redshift—the lengthening of a receding object's light waves toward the red end of the spectrum—Hubble discovered that galaxies with the greatest redshift also appeared fainter and probably more distant. Those that were farthest away were also receding the fastest—evidence of an expanding universe. The speed of recession for distant galaxies, as measured spectroscopically, thus provided a way to set intergalactic distance using what astronomers today call the Hubble constant.

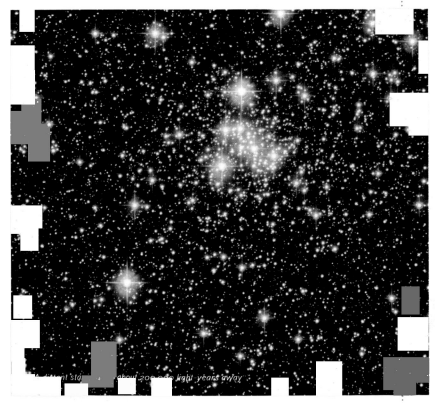

A distant star about 200,000 light-years away

YES, A CLEAR NIGHT SKY is beautiful, a glittering array of lights, but it can also be daunting to the first-time sky watcher. Surely most of those lights are stars, but are some of the planets? Is that bright object Venus, or is it an airplane? And just where are those neatly outlined portraits of dogs and bears and maidens that we see in constellation books? What follows are some basic guidelines for telling a star from a satellite, a comet from a spacecraft, and a planet from an airplane.

Stars
The twinkly, apparently motionless lights are stars. If you watch them for a while, you'll see that they do move, seeming to drift slowly overhead, rising in the east and setting in the west. They twinkle because starlight from these remote points is tossed about by our turbulent atmosphere.

The fixed patterns of the stars never change, but if you expect them to form coherent constellation pictures in the sky, you'll be disappointed. It takes time, a star map, and a good imagination to find and recognize constellations. They rarely look like anything but a scattering of stars.

Planets
The bright, steady lights are planets. Planets don't twinkle as much because the light from their relatively large disks is not as affected by our atmosphere. There are five planets visible with just the naked eye: They are Mercury, Venus, Mars, Jupiter, and Saturn.

Venus is the second brightest object in the night sky after the Moon. It shines a brilliant white and is typically best seen either in the morning or in the evening. Mars glows an orangish red, Jupiter is a steady white, and Saturn is pale yellow. You can also distinguish a planet from a star by its position on the ecliptic—the east-west path that the Sun and Moon take across the sky. All planets, because they orbit in roughly the same plane of the solar system, will be found along or near this path. They appear at different times from year to year and change position slightly from night to night.

The Milky Way & More
That blurry band of light running across the sky, high in winter and summer, is the huge collection of distant stars that make up the Milky Way, the galaxy in which our solar system is located. You'll see it best in a location far from city lights, as an urban glare can wash out its appearance. Small, blurry blobs are star clusters, nebulae, or, in the case of one little blur in the Andromeda constellation, the Andromeda galaxy, at about 2.5 million light-years the most distant object visible to the naked eye.

Fast-Movers
Some night sky objects may gently float across the sky while others are streaking quickly past. Their velocities can yield clues to their different identities and locations. Fast-moving streaks of light that look as if someone lit a match on the surface of

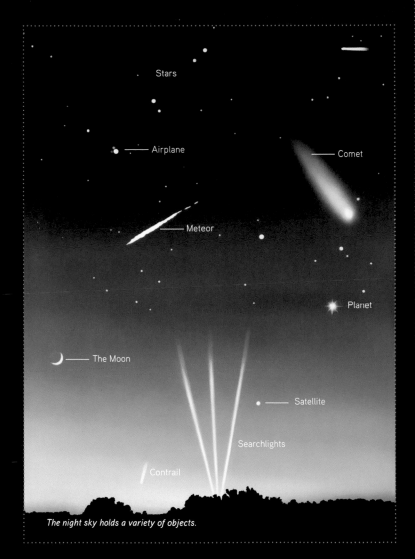

Stars

Airplane

Comet

Meteor

Planet

The Moon

Satellite

Searchlights

Contrail

The night sky holds a variety of objects.

the sky are meteors (also commonly called "shooting stars"), solar system debris flaming out in Earth's atmosphere. If the light moves more slowly, but steadily, with blinking white or steady red and green lights, that's an airplane. Just after sunset you may see smaller, starlike lights that are moving smoothly before disappearing into Earth's shadow. These gliders are satellites orbiting the Earth or, occasionally, the space shuttle.

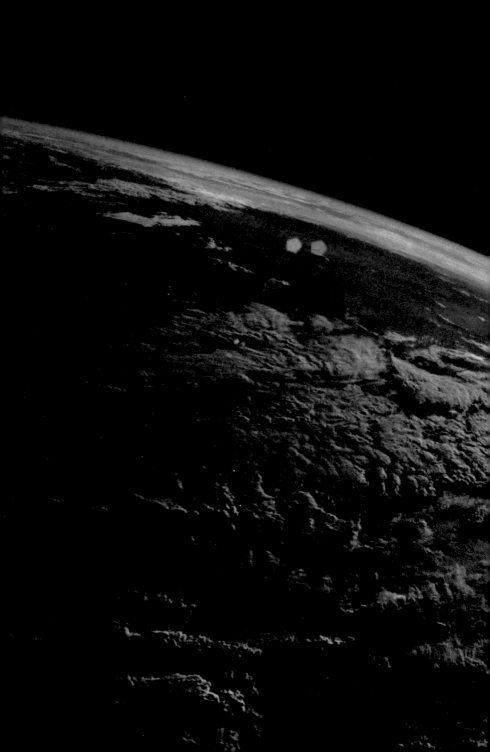

The Atmosphere

In this chapter:
*The Earth's Thin Skin
Nightfall
Earth's Magnetic Field
The Aurora*

FEATURE: *Five Coolest Things in the Sky*

• • •

Thick cloud banks populate Earth's troposphere

The Earth's Thin Skin

LIGHT FROM THE SUN and stars travel immense distances to reach Earth. Yet the last few dozen miles of the journey are what have the greatest effect on what you'll see. After crossing the comparatively empty reaches of interstellar space, the small portion of a star's light that does reach Earth must still navigate the soup of gases, liquids, and solids that make up the planet's atmosphere, which has played tricks on sky watchers since the very beginning of recorded time. As a practical matter, the atmosphere is only a few miles deep, with about 98 percent of atmospheric mass contained within some 20 miles (32 kilometers) of the planet's surface.

Inner Layers

At the lowest level, extending roughly 6 to 10 miles (10 to 16 kilometers) above Earth's surface, the troposphere has the biggest impact on visibility because weather "happens" here. Enveloping cloud systems in the troposphere can block the sky for days. Dust and pollution, particularly near cities, can overwhelm the light of all but the brightest objects.

Starting at the edge of the troposphere and extending to about 30 miles (43 kilometers), the stratosphere contains some clouds, thin wisps called nacreous clouds. The stratosphere is where upper-level ozone concentrates and blocks some of the more intense and penetrating forms of ultraviolet solar radiation.

Outer Layers

In the mesosphere, a band that runs between 30 and 50 miles (48 and 80 kilometers) above Earth, temperatures drop as low as minus 120°F. Cloud formation all but stops. Any clouds that do form—clusters of ice crystals visible—are rarely seen, and mostly at twilight in northern latitudes.

The thermosphere is the largest of the Earth's atmospheric zones as well as the hottest. Though thinly distributed, the gases present in the roughly 260-mile-wide thermosphere absorb

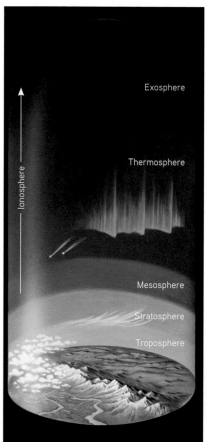

Exosphere

Thermosphere

Ionosphere

Mesosphere

Stratosphere

Troposphere

so much arriving radiation that temperatures can exceed 1800°F. The collision of solar rays and gases also produces electrically charged regions—an ionosphere—that reflect radio waves back to Earth, a benefit to modern telecommunications. The thermosphere is where the Earth's atmosphere ends, and the interstellar voyage truly begins.

THE SCIENCE OF Locating Observatories

More than a thousand years ago, Maya scientists used the raised platform of El Caracol to watch the skies. Centuries later, astronomers are still seeking the best views. The W. M. Keck Observatory, about 2.5 miles (4 kilometers) on the peak of Mauna Kea in Hawaii, is the world's highest observatory, peering from a point that is above about 40 percent of the Earth's atmosphere and 97 percent of its water vapor.

Layers of the Atmosphere

NAME	ALTITUDE	ACTIVITY
Troposphere	6 – 10 miles	Lowest level where clouds form and weather occurs
Stratosphere	10 – 30 miles	Contains ozone layer that absorbs sun's ultraviolet radiation
Mesosphere	30 – 50 miles	Temperatures drop sharply to roughly -93°C
Thermosphere	50 – 210 miles	Sun's radiation causes sharp increase in temperature, up to 1727°C
Exosphere	>210 miles	Outermost layer of atmosphere, mostly hydrogen and helium gas

Weather occurs in the troposphere.

Nightfall

THE SUN'S APPARENT motion across the sky fascinated and confused the world's finest thinkers for thousands of years. Twilight is a time to pay attention to how our closest star, our planet, its moon and atmosphere all interact.

As the Sun sets, a glance to the east can reveal a deep blue band across the horizon—the shadow of Earth, cast on the sky. In areas with high mountain peaks, a "summit shadow" may also appear. The band rises as the planet's rotation continues, until night falls and the shadow becomes invisible. Under clear skies and with a broad horizon, the pink Belt of Venus, caused by the last of the day's sunlight reflecting off the atmosphere, can be seen atop Earth's shadow.

Earthshine

Similarly, earthshine can make a crescent Moon seem almost full. Just as

Noctilucent clouds high in the sky

See also, "The Ecliptic," pp. 16-17

Backyard Guide to the Night Sky

the reflected light of the Sun puts the Moon through its phases, the Earth, to an observer on the lunar surface, would sometimes be in full shadow, sometimes partially revealed, and sometimes in full light. When a nearly "full Earth" coincides with a crescent Moon, the sunlight reflecting from our oceans and clouds and icecaps falls on the Moon, illuminating that satellite's otherwise dark portions.

Nightfall can also reveal a subtle, pyramidal glow in the western sky, visible best in latitudes closer to the Equator—and then only on darker, moonless nights. Near where the Sun sets, this is the zodiacal light, so called because it occurs along the ecliptic—the path that the Sun follows and the approximate course of the constellations of the zodiac. The zodiacal light is the result of sunlight reflected off dust in the plane of the planets' orbits left by comets and asteroids.

In more northern locations, twilight might illuminate high-flying noctilucent clouds, glowing, silver-blue wisps in the Earth's mesosphere. Far above the lower, tropospheric region where most clouds occur, noctilucent clouds form some 50 miles (80 kilometers) above Earth, a combination of ice crystals condensed around meteoric or other dust. They are a summertime phenomenon, seen when the Sun has dipped roughly 10 degrees below the horizon and almost always spotted at latitudes above 45° north—or roughly north of Boston.

Tricks of Light

The play of sunlight during the day can also produce unique effects. Rainbows—sunlight refracted and reflected into a prismatic band by raindrops—are not uncommon. Hexagonal ice crystals in high-flying cirrus clouds can bend light into a halo around the Sun or mold it into a colorful spot known as a parhelion, or "sun dog." Ice crystals can also focus the light of a low or setting Sun into a colorful pillar—similar to the beams that seem to rise from streetlamps on damp, cold nights.

THE SCIENCE OF Green Flashes

Most often seen during coastal sunsets, green flashes are rare but memorable. As the Sun sets, the atmosphere bends and disperses its light like a prism. The longer wavelengths of red, orange, and yellow bend the least and disappear first. In theory, the Sun should appear to turn green, blue, and purple in succession. But as the spectrum shortens, the light is too scattered by the atmosphere to reach Earth. Yet, when conditions are right, a green sliver can be seen on the setting Sun. Add to that the mirage effects caused by varying atmospheric temperature, and the sliver may stretch in a brief flash.

Earth's Magnetic Field

THE SUN IS EARTH'S source of energy, but solar radiation arrives at the surface only after a long journey and a long process of mediation—for which we can be thankful. The planet's atmosphere filters many potentially damaging ultraviolet and other rays, while allowing just enough energy through to keep the planet's surface temperate. But long before that, a magnetic shield, sustained far beyond Earth's atmosphere by the planet's inherent magnetism, deflects an otherwise intense atomic bombardment.

The Magnetosphere

With a swirling interior of molten metal, Earth's core generates a magnetic field that resembles a conventional magnet, but on a much larger scale. Current flows outward from the two Poles and loops back toward the center. That fact has been of practical use since the invention of the compass. But the breadth and impact of the Earth's magnetic field have become fully clear only in the satellite age, when scientists discovered the size and protective nature of a "magnetosphere" extending tens of thousands of miles into space. The magnetosphere deflects the steady stream of atomic particles known as the solar wind, a bit of "space weather" that can still knock out communications and electronic equipment during intense geomagnetic storms.

Buffeted by the solar wind, the magnetosphere changes size and shape through the day. On the side of the Earth facing the Sun, the solar wind compresses the magnetosphere down to perhaps 40,000 miles (64,374 kilometers) from the surface of the planet. On the planet's opposite side, the solar wind may "blow" the magnetosphere into a long tail of 50 to100 times that length thus extending far past the orbit of the Moon.

SKY WATCHERS: James Van Allen

James Van Allen's long-standing interest in cosmic rays led to a profound discovery. When the first U.S. satellite, Explorer 1, went into space in January 1958, it carried Geiger counters that Van Allen developed for the trip. A steady flow of evidence indicated that the Earth was encircled by two doughnut-shaped belts where the planet's magnetic field was trapping solar and cosmic radiation. Considered the first major discovery of the space age, the belts were named for the scientist.

42

Backyard Guide to the Night Sky

Earth's protective magnetic field (pink, blue, & red)

Van Allen Belts

Although the sheer strength of the magnetosphere repels some of the solar wind, atomic particles from the Sun do push their way through, as does intergalactic radiation from outside the solar system. In the late 1950s, physicists found out where some of that atomic material goes.

Girding the Earth are two concentric zones of intense radioactivity, dubbed the Van Allen belts after the physicist who discovered them in 1958. Thickest at the Equator and weaker at the Poles, the inner belt begins approximately 600 miles (966 kilometers) above the surface and

extends outward about 3,000 miles (4,828 kilometers). The outer belt runs from roughly 10,000 to 25,000 miles (14,484 to 24,140 kilometers).

The Van Allen belts are composed of atomic particles—protons and electrons from outside the solar system and helium ions from the Sun chief among them—that have pierced the outer rim of the magnetosphere, mingled with the atmosphere and, in effect, been caught in a magnetic spiral, bouncing from Pole to Pole. This accumulation of radioactive particles is so thick with energy that spacecraft navigating through them need protective shielding for their equipment.

The Aurora

THE SOLAR WIND involves a fairly constant outpouring from our Sun—roughly a million tons of matter per second, streaming away at speeds as fast as two million miles (three million kilometers) an hour. But that pales compared with the amount of material spilled during events like coronal mass ejections or solar flares, phenomena that can dump billions of tons of charged plasma and particles into space.

Polar Light Show

From Earth, these violent solar events are behind one of the most colorful and anticipated atmospheric phenomena—the dancing auroras that light up the northern and southern sky. The heavier-than-usual flow of charged

SKYFACT

Solar storms in 1989, at the peak of one 11-year cycle, caused auroras that were spotted as far south as the Caribbean.

Aurora borealis over Manitoba, Canada

particles breaches the magnetosphere and Van Allen belts and reaches into Earth's atmosphere, coming perhaps as close as 50 miles (80 kilometers) to the surface. There, the solar material energizes molecules of oxygen and nitrogen in the atmosphere and causes them to glow.

For viewers on Earth, the resulting curtains of green and red and pink light can create a dazzling spectacle, intensifying in relation to the solar storms that cause them. Data from

The aurora borealis in the polar sky has led to a number of legends among the northern-dwelling people, who welcomed—or feared—its appearance. To some, the lights were lanterns carried by spirits seeking the souls of the newly dead to escort them to the afterworld. Some Eskimo interpreted the aurora as a sign of good weather; others carried weapons for protection against it. In Norway, the lights have been linked to the legend of the Bifrost, the bridge linking Earth to the realm of the gods, and were also believed to provide contact with the dead.

satellites launched to study the auroras and other electromagnetic activity near the Earth also determined why the auroral lights seem to shimmer so vibrantly: As the solar wind blows, it actually stretches part of the Earth's magnetic field, which eventually snaps back into place. Energy released during that magnetic reconnection causes the aurora to brighten and undulate.

North and South

Auroras for the most part are polar phenomena, hence the more formal names, aurora borealis, or northern lights, when they occur in the north, and aurora australis when they occur in the south. They are seen often in the high latitudes of North America—in Alaska, certain parts of Canada, and the Scandinavian countries, where the inhabitants crafted legends around the beautiful lights. In the south they shine mostly over oceans and uninhabited land. However, when solar storms are particularly intense, they can be seen in the middle latitudes, at around 40° north and south. Solar storms tend to intensify on an 11-year cycle.

The sky is a treasure trove of noteworthy objects and events, but a number stand out. Every beginning sky watcher should seek out these ten fascinating naked-eye features.

❶ Big Dipper & Polaris

The Big Dipper in the night sky

The Big Dipper is the most familiar star pattern in the sky, and one that most people can identify with little help. Fewer people know that it is not by itself a constellation, but an asterism, a star pattern that forms part of a larger constellation—in this case, Ursa Major, the Great Bear. In the sky, the stars form a pattern resembling a ladle, or dipper. Depending on the time of year, its orientation changes; in some seasons, it appears upright and in others, upside down.

The Big Dipper is an immensely useful asterism to know. For one thing, it is visible all year round in most of the Northern Hemisphere because of its close position to the sky's north pole. (Face your north horizon to find it. It circles the sky's pole during the year, so that its dipper faces upward in the autumn and downward in the spring.) And it's a large asterism, 25 degrees long, which corresponds to the size of your outstretched fist with thumb and pinkie extended.

The asterism is also a great star-hopping guide. To star hop to the Little Dipper, draw a mental line from the stars that form the front of the Big Dipper's scoop to find Polaris, the North Star, which resides in the tail of the Little Bear.

In summer, trace a line from the stars making up the back of the dipper for about 60 degrees and spot the big, hot stars Vega and Deneb, slightly above and below that line, respectively. Or "arc to Arcturus": Follow the curve of the dipper's handle about one dipper length to the golden star Arcturus in the constellation Boötes. And then if you like, you can "speed on to Spica," following the curve onward the same distance to the brilliant blue star Spica in Virgo.

Polaris

Polaris is the North Star and a year-round "landmark" to finding your way around the constellations in the night sky. Located almost directly above Earth's northern axis, it seems to be the still point around which the sky revolves. Contrary to popular opinion, it is not the brightest star in the sky—in fact, it's the 50th brightest star—but nevertheless fairly easy to pick out if you use the Big Dipper's front stars as a guide to find it. It's part of the constellation Ursa Minor, the Lesser Bear, and marks the end of the handle of the Little Dipper asterism.

A yellow supergiant star some 430 light-years distant, Polaris interests astronomers because it is a member of a class known as Cepheid variables, stars that vary regularly in brightness. It's also part of a triple star system, although the other two stars are not visible to the naked eye.

Hikers and navigators appreciate Polaris because it marks true north—but it was not always so. Tugged by the gravity of the Sun, Moon, and other planets, Earth's axis has wobbled for over a period of 26,000 years in a motion known as precession. In the past, the Earth's North Pole has pointed toward different north stars; records from ancient Egypt indicate its astronomers used the star Thuban, a star in the constellation Draco, for northern orientation. Star charts take this into account by designating their "epoch," now typically A.D. 2000. In a few thousand years, the North Pole will move away from Polaris toward Vega, a truly bright star in the constellation Lyra, that will become Earth's new polestar.

POLARIS

Use the Big Dipper to find Polaris.

Venus's surface, seen by radar

❷ Venus

Beautiful, mysterious Venus is the brightest object in the night sky after the Moon. It shines so brilliantly, in fact, that observers often mistake it for the nearby lights of an airplane. But despite being known variously as the morning star and the evening star, it is, of course, a planet, the second from the Sun and the nearest to Earth.

The "morning" and "evening" portions of its nicknames are accurate, because Venus is visible for a few months each year either around sunset near the western horizon or around sunrise in the east. (Like all planets, the times when it appears in the sky will change from year to year, so check the planetary tables to locate it.) This is because, like Mercury, it orbits between the Earth and the Sun and so always appears near the Sun in the sky. Its interior orbit also means that Venus passes through phases, like the Moon. These are visible through a telescope. As Venus approaches Earth in its orbit, its phase becomes a thin, large crescent; as it departs, it waxes toward a full, but smaller, orb.

Venus's diamondlike brilliance is the result of two factors: its nearness to Earth and its cloud cover. At its closest point, it is only 25 million miles (40 million kilometers) from our planet, practically next door in astronomical terms. Its surface is completely obscured by highly reflective clouds at all times, and in fact 65 percent of the sunlight that reaches it is bounced back into space. This does not mean the planet is a swampy wonderland: Despite its gorgeous appearance in our skies, Venus is a hellish environment of blistering temperatures and crushing pressures.

❸ Perseid Meteor Shower

On almost any night, a patient observer might see the streak of light that marks a sporadic meteor, but during a meteor shower, "shooting stars" can flash across the sky at the rate of one or two a minute. Meteor showers occur when the Earth crosses the

Perseid tracked by time-lapse photography

debris path of a comet or asteroid; because these swarms of meteor material follow regular orbits themselves, showers occur at the same time each year (see p. 255). August brings one of the very best: the Perseids, so named because the meteors seem to originate near the constellation Perseus. At their peak on August 12, under good, moonless viewing conditions, the Perseids may pepper your sky with as many as 70 or 80 meteors an hour.

Luckily, the Perseids occur during a balmy time of the year. Best viewing times are after midnight, preferably when no Moon is in the sky; many websites can tell you the optimum times for the current year. Set your alarm, find your best dark location, and set up a comfortable lawn chair, angled back. Don't bother with binoculars or a telescope: The naked eye is best for scanning the skies. Although they come from the direction of Perseus, the meteor trails may appear all over the sky. A few may even streak directly toward you, but don't worry—they are dust-size particles that flame out even as you watch.

❹ The Pleiades

The Pleiades (M45), often known as the Seven Sisters, shine like a handful of jewels in the sky. This grouping of stars is prominent in the constellation Taurus. Follow a line from Orion's shoulder through Taurus's bright star Aldebaran and on to see the gleaming congregation of stars, highest in the sky during the fall and winter.

The Pleiades

The Pleiades are a prime example of an open cluster, a loose collection of stars bound weakly by gravity (see p. 140). Six or seven can be seen with the naked eye, and dozens are clearly visible through binoculars. This cluster actually holds about 500 stars. Located 440 light-years away, as stars go they are young, only 100 million years old or so. They will hold together as a cluster for another 250 million years before separating. Long photographic exposures show them enveloped in a misty-looking nebula because the cluster is embedded in a cloud of cosmic dust.

The seven brightest are named after seven sisters in Greek mythology. Most brilliant is Alcyone, with a magnitude of 2.86. Known since ancient times, the cluster is mentioned in the Book of Job ("Canst thou bind the sweet influences of Pleiades, or loose the bands of Orion?") and in the mythology of cultures around the world. The Kiowa of Wyoming tell the story of seven girls who were chased by giant bears onto a tall rock and then whisked into the sky. The frustrated bears clawed the rock, leaving the immense gouges that mark Devil's Tower today.

❺ Andromeda Galaxy

Look directly overhead on a fine autumn night, and you'll find a family reunion of sorts: The constellation Cassiopeia forms a distinctive W shape along the Milky Way. Surrounding the Seated Queen are her husband, Cephus, and her daughter, Andromeda. In Greek mythology, Cassopeia was a powerful queen who boasted

The Hubble Telescope's detail of the Andromeda galaxy's halo

that her daughter was more beautiful than the goddesses of Mount Olympus. Enraged, the goddesses decreed that Andromeda must be sacrificed or the kingdom would be destroyed. Chained to a rock, Andromeda was to be devoured by a sea monster; her life was saved by a hero, Perseus, who killed the beast, freed her, and saved the kingdom. According to legend, the gods placed the hero, the damsel, and her parents in the stars. Below the base of Cassopiea's W-shaped grouping, angling toward the northeast, is the plain line of stars that marks the Andromeda constellation.

In the space between the two, below the lowest part of Cassiopeia's W is a faint oval smudge. What's so great about that little spot of light? Why should we care? Because this dim 4th-magnitude blur is in fact an enormous aggregation of stars, the Andromeda galaxy (M31). It is the only galaxy visible to the naked eye in the Northern Hemisphere.

At somewhere between 2.5 and 3 million light-years away, it is by far the most distant object visible to the human eye. At an apparent magnitude of 4.4, it is one of the brightest Messier objects, but is most easily observed on moonless nights. This amazing object is one of the highlights of the night sky.

The Andromeda galaxy is spiral shaped, like the Milky Way, but it is somewhat larger than ours, containing an estimated one trillion stars. It's accompanied by several smaller galaxies, one of which, M32, can be seen through binoculars.

History & Significance

Andromeda plays an important role in astronomical history. The galaxy has been observed and studied for centuries, as astronomers slowly came to recognize both its significance as well as its beauty. The galaxy has been on record since at least A.D. 964, when it was noted by the great Persian astronomer Abd al-Rahman al-Sufi. He described the galaxy as a "small cloud" in his *Book of Fixed Stars.* The night sky object continued to generate interest over the years. Astronomer Charles Messier added the galaxy to his catalog of deep space objects in 1764.

The galaxy continued to be studied for decades to come. In the 20th century, astronomer Edwin Hubble used observations of Cepheid variable stars in Andromeda to calculate that what was then known as the Andromeda Nebula was actually far more distant than any other object in the Milky Way. His work was the first confirmation of the existence of galaxies outside of our own. This major discovery overturned previously-held ideas of the size and nature of the universe.

But to the typical sky watcher, there are several notable aspects of viewing the galaxy. It inspires notions of other worlds orbiting other stars and the possibility of life elsewhere in the universe. But perhaps the most impressive part of seeing the Andromeda galaxy is surely the idea that the light that we see now from the galaxy left home at least 2.5 million years ago—when on Earth the first members of the genus *Homo* began to walk the grasslands of Africa.

The Sun

The Sun sinks low in the sky at sunset.

Our Star

European Space Agency Solar and Heliospheric Observatory: www.esa.int/science/soho

AVERAGE SURFACE TEMPERATURE:
10,000°F (5538°C)
AVERAGE CORE TEMPERATURE:
29,000,000°F (15,600,000°C)
ROTATION: 25.4 days at Equator

DIAMETER: 924,000 mi (1,500,000 km) at
Equator
MASS: 333,000 times that of Earth
GRAVITY: 28 times that of Earth

ROUGHLY 4.6 billion years ago, in the Orion arm of the Milky Way, a spinning cloud of hydrogen and other interstellar matter succumbed to the effects of gravity and began collapsing on itself. As the cloud of gas condensed, pressure and temperature at the center increased so dramatically that the hydrogen atoms began fusing into helium, releasing immense amounts of energy in the process. Radiating outward, the energy helped counteract gravity and stop the cloud's contraction. Equilibrium set in and a star was born—our star, the Sun.

Situated about 93 million miles (150 million kilometers) from Earth at the center of what is, since Pluto's

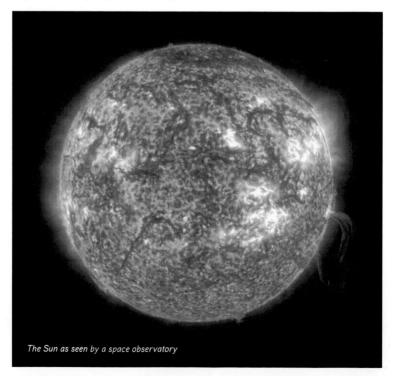

The Sun as seen by a space observatory

demotion, an eight-planet solar system, the Sun is a daily reminder of our own tenuous place in the universe. It is one of about 200 billion stars in the Milky Way, which is itself one of perhaps hundreds of billions of galaxies in the universe, and its dimensions, dynamics, and distance from Earth provide enough energy and heat to support life without overwhelming the planet with violent radiation. Had the original gas cloud been larger or smaller, or the influence of other forces different, Earth might be withering under the 860°F (460°C) surface temperatures of Venus or frozen by the minus 350°F (-212°C) conditions found on Neptune.

Composition

As it turned out, the Sun became a star of relatively modest size and heat, creating conditions on Earth temperate enough to let the oceans form while delivering a steady supply of about 1,400 watts of energy to each square yard of the planet. The Sun's roughly 11,000°F (6093°C) surface gives it a yellowish hue and makes it a G2-type star, near the middle of the stellar classification scheme for color and temperature, though a bit brighter and hotter than most other stars in the Milky Way. (For more on star classification, see pp. 132-133.) A further rating marks it as a yellow dwarf, lying amid what astronomers call the main sequence of stellar life cycles.

Like all stars, the Sun is primarily made of hydrogen, the fuel for the atomic reactions at its core. Hydrogen represents just over 92 percent of the Sun's matter, with helium making up just over 7 percent, and trace amounts of sodium, iron, and other elements accounting for the small remainder. What we see from Earth—the layer known as the photosphere—has substantially more helium, about 25 percent, with about 74 percent hydrogen and small amounts of oxygen, carbon, iron, sulfur, and many other elements.

In the context of other stellar objects, the Sun is unremarkable. Yet it has fascinated humans for centuries. It was treated as a god by the ancient Egyptians. Eclipses have figured in tales as diverse as the Chinese legend of the "devouring dragon" (local residents would bang pots to scare the beast away). The Sun weaves so deeply into our daily life that it not only governs external conditions like the weather but also triggers our internal synthesis of vitamin D and can sway human psychology as well.

THE SCIENCE OF Living Dangerously

Living, as Earth does, even in rough balance with a star is inherently treacherous. The Sun's radiation output runs from high-energy and life-damaging gamma rays, x-rays, and ultraviolet rays. Much of the more dangerous stuff is buffered by the Earth's atmosphere and magnetic field. Still, depletion of the planet's upper ozone layer has been connected with rising skin cancer rates as more ultraviolet light reaches Earth's surface. Warnings about Sun exposure and the need for protective clothing or sunscreen have increased in proportion.

The Life of the Sun

AVERAGE IN its structure compared with other stars, the Sun has an average life expectancy, too—about 11 billion years, which puts our solar system squarely in its middle age. A smaller, cooler-burning star, such as a red dwarf, would live for much longer time, perhaps into the tens or even hundreds of billions of years, whereas a hot supergiant star might spend its fuel in a million years or so. Our star probably has about five billion years of primary fuel remaining, and another few hundred million years before it fades from the sky.

That is still an unimaginable amount of material. Each second, nuclear fusion in the Sun consumes about 700 million tons of hydrogen, creating 695 million tons of helium in the process. The excess 5 million tons of matter are converted directly into energy. But one day the tank will run dry. What happens then?

Beginning of the End

The endgame of a star's life depends on how it lived—primarily on its size and temperature. In the case of our yellow dwarf Sun, as it begins to run out of its chemical fuel, the amount of

SKYFACT

The Sun accounts for about 99.8 percent of the solar system's mass, and like the rest of the universe, is made almost entirely of hydrogen and helium. It does contain trace amounts of other elements, including iron, nickel, oxygen, silicon, sulfur, magnesium, carbon, neon, calcium, and chromium.

An imagined view of Earth's barren surface under a red giant Sun

See "Life Cycles of Stars," pp. 136-137

Backyard Guide to the Night Sky

energy pouring outward would begin to decline. The equilibrium established billions of years ago between outward-moving energy and the inward pull of gravity would be upset. As gravity continued to assert itself, the Sun would begin to collapse.

That contraction would raise the Sun's inner temperature and give the star a last gasp of brilliance. Leftover hydrogen, away from the Sun's fusion-furnace core, would begin to burn, and the helium created from billions of years of nuclear reaction would also ignite, fusing into carbon.

Death of the Sun

The energy from these new reactions would cause the Sun to expand far beyond its current boundaries as it enlarged into a red giant. This engorgement would lead to the Sun's

engulfing an area likely to include the current orbits of Mercury and Venus—and perhaps swallowing both planets as well, along the way. The Sun by this point would have boiled away the Earth's oceans and all of its inhabitants. The planet Earth itself, as well as others in the solar system, would move outward into farther orbit as the Sun spewed its gas and matter into space and its gravitational hold continued to weaken. The star might begin to pulse, turning into a variable star. In the next phase, ejected gases would form a cloud—a nebula—as they began to dissipate.

Once the second round of fuel consumption had come to an end, the fusion reaction would halt altogether, and gravity would again begin pulling the Sun's matter into itself. The remaining material, including what was left of the core, would be compressed into an area about the current size of the Earth as the Sun became a hot, but gradually cooling, white dwarf—drifting remnant of a once unique corner of the universe.

57

Anatomy of the Sun

FROM OUR vantage point the Sun seems placid. Though its heat can be intense, the view from Earth is of an object that shines as it moves through the sky. But as spaceborne probes and scientific instruments have become more sophisticated, our understanding of the Sun shows it to be a roiling, complicated place of violent extremes, with its own weather systems and an almost disturbing hint of unpredictability.

The whole star rotates, but in an eccentric way: The equatorial region of the Sun moves as if it is a solid, completing a turn in 25 or so of our days, whereas the polar regions lag behind and take 34 days or so for one

rotation. Along with the regular flow of energy, phenomena like solar flares and coronal mass ejections send billions of tons of atomic material hurtling toward Earth—enough to affect our communications and electrical systems—and, potentially, our health.

Anatomy

The heart of the Sun is its dense, high-pressure core, where roughly half of the star's mass has been squeezed into about 7 percent of its volume. It is here that the process of nuclear fusion takes place. Temperatures can reach an astounding 27,000,000°F (15,000,000°C). As hydrogen is converted to helium, packets of energy called photons are emitted and begin a journey to the surface that can take millions of years to complete.

THE SCIENCE OF Sunquakes

Since the 1960s, several solar probes have been launched to study our nearest star but have gone no closer than about 28 million miles (45 million kilometers). To unravel what is happening inside, scientists have to rely on a variety of indirect methods, one of the most productive being helioseismology—the study of how low-frequency sound waves move through the Sun's various regions. Just as scientists on Earth study the shock waves created by earthquakes, helioseismologists study sunquakes—the waves generated inside the Sun by the convection currents that transmit heat and solar material from the center of the star to the surface. As they work their way outward, these waves cause the Sun's surface to heave up and down. By studying the progress and effect of these waves on the Sun's surface, scientists can develop a picture of what is happening underneath.

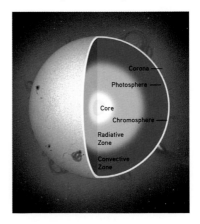

A solar flare erupts.

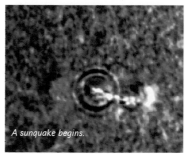
A sunquake begins.

Seismic waves spread.

Waves of a solarquake ripple.

Fighting their way from the core, the photons begin losing energy and cooling. About three-fourths of the way to the surface, they reach an area where convection currents continue drawing gas and heat outward. They then reach the photosphere—a roughly 300-mile-thick (483 kilometers) layer. At this point, temperatures have fallen to around 11,000°F (6093°C).

The brilliant, steady light emitted at the photosphere makes the Sun appear as a solid, not as a turbulent and multilayered surface. Photos of the Sun make it appear grainy and in rapid boil, with bubbles of gas pushing to the surface, settling out and sinking back into the stew over about ten minutes' time. A pinkish layer, called the chromosphere, surrounds the outer surface of the Sun, firing out spicules, or spikes, of gas. It is visible only when the Sun is in total eclipse (and when special equipment is used to protect the naked eye). Solar prominences—clouds of gas—float across the surface or shoot out in arcing geysers.

Surrounding that, the Sun's far-reaching corona drifts for millions of miles—a ghostly and, except during total solar eclipse, invisible halo of gas where temperatures inexplicably rise back to perhaps 2,000,000°F (1,111,093°C). Compared with the dense solar core, the corona is almost nonexistent—trillions of times less dense than the air on Earth. Holes in the corona, caused by the Sun's magnetic field, are where steady streams of particles known as the solar wind break free and flow toward the rest of the solar system.

Sunspots & Solar Flares

I N THE MID-19th century, amateur astronomer Richard Carrington determined that patchy spots near the Sun's equator rotated around more quickly than did spots located farther from the middle. In 1859 he also noted—as did amateur Sun watcher Richard Hodgson—the eruption of a patch of white light near a sunspot group, an event that coincided with telegraph system problems as well as a brilliant display of auroras. The sig- nificance of those observations was not clear at the time. In fact, the phe- nomena involved remain under study even today among scientists trying to determine the interplay between the Sun's weather and its magnetic field.

The Sun's Rotation

Carrington's observations were explored more fully by the European Space Agency's Solar and Helio- spheric Observatory (SOHO). In the

Coronal loops arc over the Sun's surface.

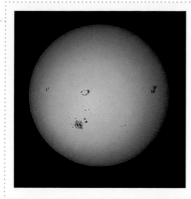

mid-1990s, data from SOHO showed that the Sun's dense but still gaseous center rotated more like a solid, while the outer areas, and particularly those near the Sun's poles, rotated much more slowly. That differential rotation stretches the Sun's magnetic field lines, which get tangled even further by the continuing surge of gas through its convection zone. Those disrupted currents can slow the flow of solar material, causing it to cool and darken in color. Appearing as a dark splotch on the surface, sunspots can be as much as 3600°F (1982°C) cooler than the Sun's typical surface temperatures of around 11,000°F (6093°C).

Sunspots

Typically clustered near the Sun's equator, sunspots peak and ebb in an 11-year cycle that coincides with other solar activity. The buildup of magnetic tension eventually gives way, and its release can eject billions of tons of atomic particles in a solar flare. The cause of the auroras that can be seen on Earth, these heavier than usual doses of radiation can knock out communications equipment and satellites and disrupt the flow of electricity in terrestrial power grids.

Because of the implications for earthbound technology and systems, predicting solar storms has become a priority for agencies like the National Aeronautics and Space Administration (NASA). NASA has upcoming plans to launch the Solar Dynamics Observatory (SDO). The SDO aims to connect information about the movement of charged plasma in the Sun's interior with changes in the magnetic field closer to the surface—and ultimately try to develop models to predict when solar storms are gathering to erupt.

The study of the Sun's magnetic field extends beyond the solar surface. The magnetic field influences activity in the Sun's corona and may be responsible for the unusually high temperatures found there. Indeed, though the Sun's magnetic field is considered weak overall, its influence is felt far into space. The so-called heliosphere extends perhaps one hundred times the distance between the Earth and the Sun (100 AU), providing a sort of curtain between our solar system and the incoming interstellar wind.

The Sun's Path

THE SUN'S APPARENT journey through the sky—and the changes to its route over the course of a year—provided an early sense of how Earth's position in the universe affects daily life. Of course it took centuries to sort out the fact that the Earth is doing the moving. Still, the daily change in the relative position of the two bodies has long found practical application, providing a way to tell time, judge the seasons, and mark annual celebrations.

The Sun's path through the sky, a line known as the ecliptic (the word also refers to the Earth's route around the Sun), was engraved by the origins of our solar system. As the Sun's birth cloud began to spin and take shape, it flattened into a disk. With hydrogen settling at the core, other material collected farther out and coalesced into what would become the planets. These planetary bodies would begin to orbit in roughly in the same plane as the original solar cloud.

Earth, as we know, moves in two distinct ways—rotating on its axis while also progressing steadily in its annual trip around the Sun. The planet's daily rotation moves in what from our vantage point is west to east. That makes the Sun and stars appear to move in the opposite direction, from east to west, in our sky.

Changing Seasons

The orbit around the Sun, meanwhile, is marked by Earth's axis of rotation, which is tilted about 23.5 degrees off perpendicular from the ecliptic. That tilt accounts for the seasons, and also for the lengths of time that the Sun appears in the sky throughout the year.

On the first day of summer in the Northern Hemisphere, typically June 21, the northern half of the planet is tipped toward the Sun. Near the North Pole is a season of 24-hour daylight. Farther south, night occurs, but the day is the longest of the year, with the Sun at its highest point in the sky.

Each day for the next six months the Sun's path will flatten a bit and the days will shorten, as autumn sets in and winter approaches.

THE SCIENCE OF Seasons

The geometry of Sun and Earth yield four noteworthy dates each year. The winter and summer solstices typically occur on June 21 and December 21—the days with the least or the most sunlight, depending on whether you are north or south of the Equator. Technically, these are the days when the Sun reaches a point directly over the Tropic of Cancer (23° 27' north) or the Tropic of Capricorn (23° 27' south). It also marks the Sun's farthest distance from the celestial equator, an imaginary circle drawn outward from the Earth's Equator. On March 21 and September 22 respectively, the spring and autumn equinoxes are when the Sun at noon is directly over the Equator—which also happens to be the two days when the ecliptic and the celestial equator intersect. This time is when the division of night and day is about the same around the globe.

See "The Ecliptic," pp.16-17

Backyard Guide to the Night Sky

Positions of the Sun

EVENT	DATE	SUN'S POSITION
Spring Equinox	03/21	Equator (0°)
Summer Solstice	06/21	Tropic of Cancer (23° 27' N)
Autumn Equinox	09/22	Equator (0°)
Winter Solstice	12/21	Tropic of Capricorn (23° 27' S)

Below the Equator, the same dynamic is at work in the south—with the dates reversed. June 21 marks the shortest day of the year, with that part of the Earth tipped away from the Sun. The equatorial areas, of course, experience none of these extremes. Located around the Earth's middle, they have days and nights that stay roughly the same throughout the year, with the seasons offering little variation.

SKYFACT People like the Pueblo Indians in the southwestern United States developed systems for tracking events like solstices and equinoxes. Places like Fajada Butte, in Chaco Canyon, New Mexico, revealed diagrams and etchings aligned to mark those key points in the solar and lunar cycle.

Time-lapse photography reveals the Sun's path as it sets.

Viewing the Sun

NASA's tips of safe Sun observation: eclipse.gsfc.nasa.gov/SEhelp/safety.html

WALK OUTSIDE on a sunny day, and the natural reaction is to shield your eyes from the intense light. It is a well-founded protective instinct: A direct view of the Sun can damage the naked eye. Channel that light through a telescope or binoculars, and the effect can be quick and catastrophic: blindness.

It is unfortunate that the closest star is also the one celestial object that requires safety precautions to view in detail. But the point needs emphasizing, particularly if kids are involved in your viewing. Caution should be the watchword, lest anyone be tempted to flash the binoculars skyward or move under the telescope to "take a peek."

Solar filters render a safer view.

Safe Observation

Some simple rules and a bit of extra equipment will let you view the Sun safely—watching its granular surface bubble, tracking sunspots, and perhaps catching a glimpse of an erupting solar flare. These are events that can be seen during the day—the earlier the better, preferably from a grassy area or across a body of water, where the air will remain cooler and more settled—and are impervious to the light pollution around cities.

If you are not using binoculars or a telescope, safe solar viewing can be done through a plate of number 14 welder's glass. With this protection, you can hunt for sunspots with the naked eye.

Through a Telescope

If you are using equipment but don't expect to do regular solar viewing, the cheapest way would be to use your telescope (or tripod-mounted binoculars) as a makeshift projector. To orient your equipment, use its shadow as a reference point, adjusting until the shadow is at its smallest. Again, do not look through the equipment for that initial orientation—and if your telescope has a finder attached, make sure to cover the aperture so no one can look through it. Using a stand or clip, affix a sheet of white paper or card-

SKYFACT

You don't need an advanced degree in astrophysics to make important discoveries about the Sun. Some of the most important findings about the Sun, including the sunspot cycle and the connection between sunspots and solar storms, have been made by amateur astronomers.

board a foot or so behind the eyepiece. You'll see the Sun's projection and can focus to sharpen the view. Cut a hole in the middle of a piece of cardboard to fix onto the barrel of the scope in a way that shields the "screen" from ambient light.

When using this method, remember to protect your equipment. Heat from the Sun can build up inside the telescope tube and damage it. A telescope aperture should be covered with cardboard, or other shield, with a hole that reduces the open area to no more than 4 inches (100 millimeters)—and going as small as 2 inches (60 millimeters) isn't being too cautious.

For more serious viewing, the easiest solution is to invest in a solar filter, a piece of metal-coated glass or plastic that will cut the Sun's brightness by a factor of perhaps 100,000. (Never use a solar filter that only fits over the eyepiece; these crack without warning under the solar heat, destroying the observer's eye.) Less expensive filters are made of aluminum-coated Mylar and can even be hand-cut from sheets of the stuff. Glass filters will provide more realistic colors and can be bought in sizes to fit the aperture of your telescope.

For even more intense (and expensive) solar viewing, a high-end H-alpha filter will restrict what you see to the narrow band of light emitted by hydrogen, the Sun's dominant gas. By eliminating the rest of the spectrum, you can see the Sun in more detail, and events like wispy solar prominences.

A N ECLIPSE is a breathtaking and spooky sight. The darkness at noon experience of a total solar eclipse, in particular, is well worth seeking out. But even the more commonly seen lunar eclipse is a fine sight and a good lesson in celestial mechanics.

What Are Eclipses?

An eclipse happens when either the Moon or the Earth blocks the light from the Sun. In a lunar eclipse, the Earth lies directly between the Sun and the Moon, so that the Earth's shadow falls on the Moon. This, of course, can happen only during a full Moon, when the Moon is on the opposite side of the Earth from the Sun. In a solar eclipse, the roles are reversed: The Moon comes between the Sun and the Earth, and the shadow of the Moon falls on the Earth. This can happen only during a new Moon.

You'd think that each kind of eclipse would happen once a month, but in fact each occurs about twice a year. This is because the Sun, Moon, and Earth do not orbit in exactly the same plane. The Moon's orbit is tilted about five degrees from that of the Earth's, so eclipses are possible only when the plane of its orbit intersects the plane of the Earth's orbit. In any year, there are between two and seven eclipses, solar and lunar combined.

The Sun's corona is revealed during a solar eclipse.

When the Sun shines on the Moon or Earth, the resulting shadow has two parts: a narrowing inner "umbra," where light is completely blocked, and a widening outer "penumbra," where light is only partially blocked. Total eclipses occur within the umbra, and partial eclipses within the penumbra.

During a lunar eclipse, the Earth's shadow envelopes the Moon, and the darkened Moon can be seen from anywhere on the nighttime side of our planet (if the sky is clear). But during a solar eclipse, the Moon casts a relatively small shadow on the much bigger Earth, so a total solar eclipse can be seen only along a small band on the turning Earth, an area that changes from eclipse to eclipse. It's important to know your location when planning to view an eclipse for this very reason.

A solar eclipse is a marvelous thing to see, but never look at the Sun directly, or you risk serious eye damage (sunglasses don't provide enough protection). View the eclipse indirectly, through projection, or using approved solar or number 14 welder's glass.

A full Moon at the beginning of a lunar eclipse

Solar Eclipses

SOLAR ECLIPSES come in three varieties: annular, partial, and total. Annular eclipses occur when the Moon is too far from the Earth to completely cover the Sun; its inner shadow, the umbra, falls short of Earth's surface. During an annular eclipse, the Moon's disk cuts into the Sun and eventually crosses its face, but the Sun's bright edge is visible all around it, forming a ring or annulus (hence the name). Your surroundings will grow dim but not dark.

How a Solar Eclipse Works

Partial solar eclipses take place when your viewing site lies within the Moon's outer shadow, its penumbra. They are typically undramatic, dimming the sunlight only a bit, but with proper eye protection you can see a bite taken out of the Sun.

The big kahuna of eclipses is the total solar eclipse. These are visible only within the Moon's umbra, when from the ground the Moon's disk completely covers the Sun. The path traced by this dark inner shadow is typically about 200 miles (300 kilometers) wide, and often narrower. The slender trail of a total solar eclipse varies from year to year in an 18-year pattern. In August 2008, for instance, it crossed the Arctic, Greenland, and Russia, and in 2010 it will curve across the South Pacific. The next one to cross the United States will take place in 2017.

A total solar eclipse is an extraordinary experience. During totality—when you are in the central shadow of the

Moon—daylight fades to dark twilight. Birds roost, crickets chirp, and stars appear in the sky. Sky watchers can see the Sun's spectacular corona, its fiery outer atmosphere, feathering out around the edge of the Moon's disk. It is safe to look at the eclipse with the naked eye only during totality, but if there's any uneclipsed Sun visible, you must use a proper solar filter.

Many cultures have understandably feared solar eclipses and believed they were signs of bad fortune. The ancient Chinese, seeing such an eclipse, would bang drums to drive away the dragon eating the Sun—a technique that invariably succeeded.

The Moon's shadow falls on Earth during an eclipse.

Umbra

Penumbra

The geometry of a solar eclipse

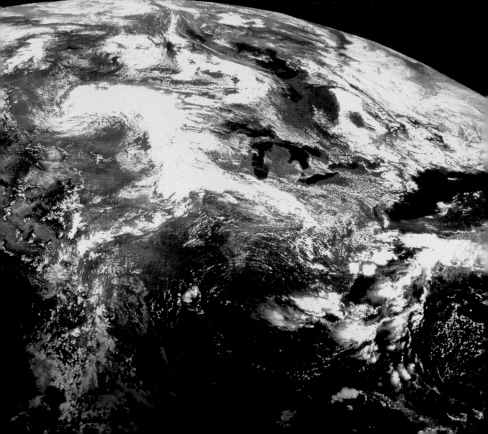

Lunar Eclipses

L IKE THE SUN, the Moon experiences three kinds of eclipses. If it passes through the penumbra—the outer region of Earth's shadow—we see a penumbral lunar eclipse. This can be hard to detect, because the Moon will typically grow only a bit dimmer than usual.

If just some of the Moon passes through the umbra, Earth's dark inner shadow, then the Moon undergoes a partial lunar eclipse. The Earth's umbra carves a passing bite out of the Moon's disk.

When the Moon moves completely into Earth's umbra, we see a total lunar eclipse. However, even then the Moon will not become completely dark, like the Sun during a total solar eclipse. During the eclipse, sunlight bends through Earth's atmosphere and shines dimly onto the Moon's surface. This refracted sunlight is typically red, the color of sunset or sunrise on Earth, so the darkened Moon will turn a rusty or bloodred hue. If the Earth's atmosphere is unusually dusty, blocking more of the sunlight, the Moon may grow quite dark and barely visible. At other times, it may stay fairly bright and orange.

Lunar eclipses can last as long as an hour and a half or pass by in a few minutes. You can view them safely with the naked eye. Binoculars are helpful; with a telescope, use a low-power eyepiece so you can see the whole Moon at once.

Sky watchers have tracked lunar eclipses since ancient times; both the Chinese and the Maya, for instance, kept careful tables of such occurrences. In 1504, Christopher Columbus, using his astronomical tables, famously tricked the inhabitants of Jamaica. Stranded on the island and needing food from the native Arawak (who were tired of their mistreatment by the European sailors), Columbus consulted his almanacs and noted that a total lunar eclipse was due. He told the Arawak that God was displeased with them and would darken the Moon with his wrath. Sure enough, the Moon turned the characteristic bloody red of a lunar eclipse. The locals promised complete aid if only God would relent, which he appeared to do about an hour later.

The mechanics of a lunar eclipse

Earth's rust-colored shadow

Penumbra

Umbra

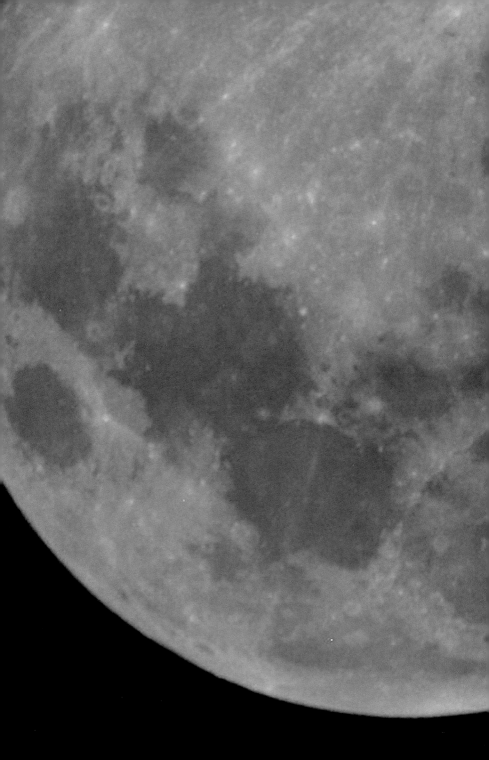

The Moon

The Moon's craters and seas in full view

Our Satellite

RADIUS: **1,079.6 mi (1,737.4 km)**
MASS: **7.3483 x 10²² kg**
DISTANCE FROM EARTH: **238,855 mi**
(384,400 km)

ORBITAL PERIOD: **29.53 days**
ROTATIONAL PERIOD: **27.32 days**
ORBITAL CIRCUMFERENCE: **1,423,000 mi**
(2,290,000 km)

THE MOON IS a great friend to stargazers, particularly novices. It is an object that is easy to find as it outshines even the worst light pollution. While easily viewed with the naked eye, binoculars will bring out many details and amazing features on the Moon.

It orbits at an average distance of about 239,000 miles (384,400 kilometers). Its 2,159-mile (3,475-kilometer) diameter is roughly one-fourth that of Earth's, but its composition from lighter elements gives it about an eightieth of the mass and about one-sixth the gravity of Earth.

A massive collision ejected Earthly debris that became the Moon.

Backyard Guide to the Night Sky

Waters recede at low tide.

Another Moon origin theory: It formed as a separate object, only to be trapped by Earth's gravitational pull.

Lunar Views

The Moon's power is felt in the daily ebb and flow of tides. Bodies of water facing the Moon are pulled toward it, causing a high tide on one section of Earth, while a second high tide is caused on the other side by the effect of Earth being pulled away from the water.

The satellite's topography is varied. Collisions with meteors over its 4.5-billion-year life have created a surface of pulverized rock, called regolith. Craters are as wide as 1,600 miles (2,575 kilometers) across, with mountainous walls as high as 4.8 miles (7.7 kilometers). Its distinctive "seas"—so called because early observers thought they might be the beds of dried-up oceans—are actually areas given a smooth sheen by molten lava brought to the surface after major impacts that occurred around 3.8 billion to 3.9 billion years ago.

Lunar Origins

Evidence from lunar missions supports the idea that the Moon formed from the debris ejected when an object about the size of Mars struck the young Earth. That would account for the presence of rocks similar to those found near the surface of the Earth, the lack of water (vaporized during the explosion, though trace amounts have been found in some lunar rocks), and aspects of its orbit (the spin of the two bodies undercuts other lunar-birth theories).

75

The Moon in Motion

THE FACT that the Moon keeps the same side exposed to the Earth might make it appear a somewhat passive traveler—trapped in an orbit and asserting little influence. The relationship between the Earth and its Moon, however, is a bit more complicated, with each influencing the other in a complex partnership.

Rotations

The Moon is, in fact, spinning on its axis, just as Earth does. But in the case of the satellite, its rate of rotation matches the rate of its progress around Earth—it takes about 27.3 Earth days to complete both. This came about because, as the two bodies evolved so close to each other, Earth's gravity created a bulging "land tide" in the Moon that caused its spin to slow and synchronize with the speed of its orbit. For earthbound observers, that means only one side of the Moon is visible, while the other side faces away. It takes

a bit longer, about 29.5 days, to complete what is known as the Moon's synodic month—the time needed to orbit the Earth and return to the same position relative to the sun, completing a full cycle of lunar phases.

Often referred to in error as the "dark side," the half of the Moon facing away from Earth is not in constant darkness. It experiences phases just like the side facing Earth, with a time each month when it is in full sunlight and a time when it is facing away from the sun and in full shadow. The Moon's far side has been mapped through space-based observation, beginning in 1959 when Russia's Luna 3 spacecraft took the first pictures of it, and followed by detailed maps created over three decades of U.S. missions. The Russian feat is reflected in some of the names attached to the Moon's far-side features—including the Moscow Sea and a crater named in honor of Russian cosmonaut Yuri Gagarin, the first human to orbit Earth. Unlike the visible side, the far side is dominated by craters with few seas, the result of its thicker crust.

THE SCIENCE OF Rotational Energy

Friction from interations of the Moon and Earth are slightly slowing Earth's rotation. But, this lost energy does not just disappear: Between orbiting celestial bodies, "angular momentum" also must be maintained. A product of mass, velocity, and the distance between two objects, angular momentum reflects the force of the spinning motion between the two bodies. If one of the objects slows—like the Earth—one of the other variables must change to accommodate, in this case the distance between them. In fact, the Moon pulls away from the Earth by about 3.8 centimeters—an inch and a half—each year.

Orbits

It is a bit misleading even to say that the Moon orbits Earth. Though the Moon's gravity is only about one-sixth that of the planet's, the two in essence

The Moon's phases during its orbit

orbit each other. Think of it as a spinning baton—albeit a lopsided one. Because the Earth is so much larger and denser, the balancing point of the system—known as the barycenter—lies about 1,100 miles (1,770 kilometers) below Earth's surface. This center of gravity for the Earth-Moon system is what actually tracks the ecliptic plane around the Sun, with the rest of the planet wobbling around it.

The fact that the Moon has an appreciable gravitational tug proved important in planning the manned Apollo missions. The flights followed a trajectory that placed them ahead of the Moon, and they relied on its gravity to pull them around to the far side. By reducing speed, the spacecraft were captured in lunar orbit. In the case of Apollo 13, the mechanics of trajectory were used to save the crew's life. After an oxygen tank exploded about 200,000 miles (321,869 kilometers) from Earth, backup power systems were used to boost the craft into a free-return trajectory in which Earth's gravity and Kepler's laws of motion caused it to loop around the Moon and head home.

The far side of the Moon

Lunar Movements

JUST AS EARTH'S progress in its orbit changes the Sun's apparent path in the sky each day, the Moon's trip around the Earth determines where it rises and sets, its course, and how much of its surface can be seen. It takes just over 29 days for the Moon to pass once around the Earth, which translates into a daily shift of about 12 degrees: Track the position of moonrise and you'll see it shifting toward the east in the sky by about that much each night.

Location in the Sky

The Moon's orientation to the ecliptic also determines its height and course. The Moon is offset from the ecliptic by around five degrees and thus holds close to that plane. In a Northern Hemisphere winter, the ecliptic is low in the daytime sky but arcs high and

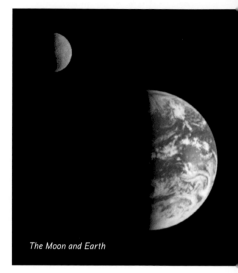

The Moon and Earth

long through the night. The reverse is true in summer, when it tends to stay low in the sky. During spring the ecliptic is angled sharply in the western sky during the early evening, while in autumn the situation is reversed, with the plane angling upward in the east around dawn. These two times of year offer the best chance to spot and study waxing and waning crescents.

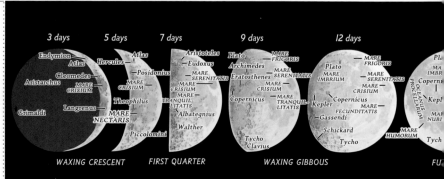

3 days — Endymion, Atlas, Cleomedes, Aristarchus, Grimaldi, Hercules, Atlas, Langrenus, MARE CRISIUM, MARE NECTARIS

5 days — Atlas, Posidonius, MARE CRISIUM, Theophilus

7 days — Aristoteles, Eudoxus, MARE SERENITATIS, MARE CRISIUM, MARE TRANQUIL-LITATIS, Albategnius, Walther, Piccolomini

9 days — Plato, MARE FRIGORIS, Archimedes, Eratosthenes, MARE SERENITATIS, MARE CRISIUM, Copernicus, MARE TRANQUIL-LITATIS, Tycho, Clavius

12 days — Plato, MARE FRIGORIS, MARE IMBRIUM, MARE SERENITATIS, MARE CRISIUM, Copernicus, Kepler, MARE FECUNDITATIS, Gassendi, Schickard, MARE HUMORUM, Tycho

Pla, MA IMBR, OCEANUS PROCELLARUM, Coperni, Kepl, MAI NUBI, Tych, FU.

WAXING CRESCENT **FIRST QUARTER** **WAXING GIBBOUS** FU.

Phases of the Moon

The Moon generates no light of its own; moonlight is reflected light from the Sun, which, in turn, causes the moon's different visual phases. As the Moon travels, the angle formed by the Sun, the Moon, and the Earth constantly changes. At one extreme the Moon lies between the Sun and the Earth; the fully backlit body, called a new Moon, cannot be seen (unless the alignment produces a solar eclipse). At the other, the Moon and Sun sit on opposite sides of Earth, and a full Moon occurs (if the alignment is right, a lunar eclipse may occur).

In between, the new Moon moves gradually along its orbit and begins to capture a few degrees of sunlight. Within a day or two, a thin waxing crescent will be visible. After a week, the Moon will form a roughly 90-degree angle with the Earth and the Sun, and about half of it will be lit—a phase called the first quarter Moon because one-fourth of the lunar cycle has passed. Next, it becomes a gibbous Moon, from the Latin word for

"hump," before becoming full. Swinging around the Earth's other side, the Moon begins to shrink, first through a waning gibbous phase, then to its half-lit, last quarter phase, then to a waning crescent, until it disappears into the dark and the next cycle begins.

There is a direct connection between lunar phases and the Moon's patterns of rising and setting. Because of the geometry involved, for example, a Full Moon will rise with the setting Sun, since they are directly opposite each other, and set only with the following day's sunrise.

17 days — 20 days — 22 days — 24 days — 26 days

WANING GIBBOUS — LAST QUARTER — WANING CRESCENT

Face of the Moon

THE MOON has led a violent life, particularly in the first half-billion years or so, when debris remaining from its formation regularly bombarded the surface.

Smooth Seas

In the 1600s Galileo observed many of the Moon's ridges and craters through a telescope. Calm and flat areas continued to be called maria (pronounced MAH-ree-uh), the Latin word for "seas." The Latin names endure today, the most famous being Mare Tranquilitatis, the Sea of Tranquility, where Neil Armstrong took his first giant steps.

The smooth maria are a clue to the Moon's history. Early on, it suffered repeated major impacts, with the larger pieces of debris left over from its formation—collisions that produced holes that were hundreds

Manned Missions to the Moon

MISSION	LAUNCH DATE	NOTABLE EVENT
Apollo 11	07/16/1969	First manned lunar landing
Apollo 12	11/14/1969	Exploration of Ocean of Storms
Apollo 14	01/31/1971	Photography and exploration
Apollo 15	07/26/1971	First use of lunar rover
Apollo 16	04/16/1972	Surveying and sampling of Descartes region
Apollo 17	12/07/1972	Last lunar flight. Surveyed Taurus-Littrow region

A moonscape as captured by an Apollo astronaut

The illusion of a gigantic full Moon rising near the horizon only to appear its usual size once it rises fully in the sky is well documented. The phenomenon has not been explained fully. One theory has been tested, and results have shown that for some reason (also not fully clear) looking up causes people to perceive objects as smaller, while seeing them on a direct horizon makes them appear larger.

of miles wide but relatively shallow at only 10 miles (16 kilometers) deep. These massive impacts allowed dark, iron-rich lava from the Moon's core to pour out, filling the holes and leaving a comparatively smooth surface. Both the average size of the objects hitting the Moon and the frequency of impacts have decreased markedly over time, leaving these newly resurfaced areas comparatively intact.

Craters

Elsewhere on the Moon's surface, collisions simply gouged holes in the surface, creating areas that are now saturated with craters. They are the areas that appear light when seen from Earth. The last major crash has been dated to about 109 million years ago, creating the crater now known as Tycho, after astronomer Tycho Brahe. Individual craters can be dozens of miles wide and several miles high, with ray-like sprays of rock sometimes emanating from their base as evidence of the impact that caused them.

The first and third quarter phases— when sunlight is striking at an angle— bring out details in sharp relief. Prominent craters like Copernicus and Theophilus can be studied with binoculars. With a telescope, the maria can be used as a reference to pinpoint craters and other objects using the Moon chart that follows.

The Moon's Near Side

Moon Features

The Moon's synchronous orbit with Earth—spinning on its axis in the same time it takes to move around the planet—means only one side is visible to land-based observers. The visible side includes lighter colored, cratered highlands, where collisions with meteors and comets have left dozens of broad, deep holes. The large, dark maria are places where molten rock has oozed out and filled large impact basins, leaving a smoother surface. Seas and craters are labeled on the accompanying chart for your reference.

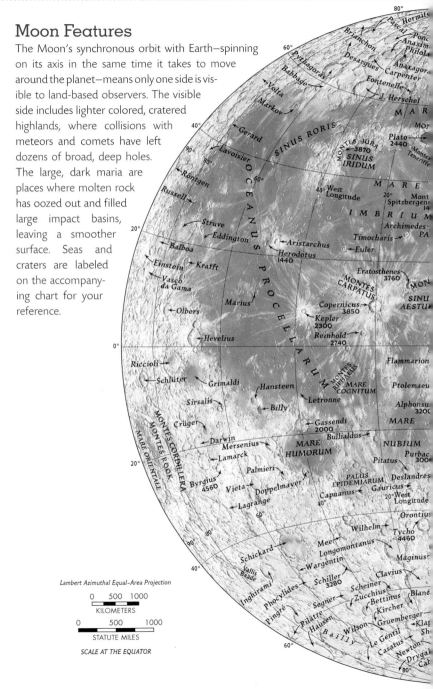

Lambert Azimuthal Equal-Area Projection

0   500   1000
KILOMETERS

0   500   1000
STATUTE MILES

SCALE AT THE EQUATOR

Backyard Guide to the Night Sky

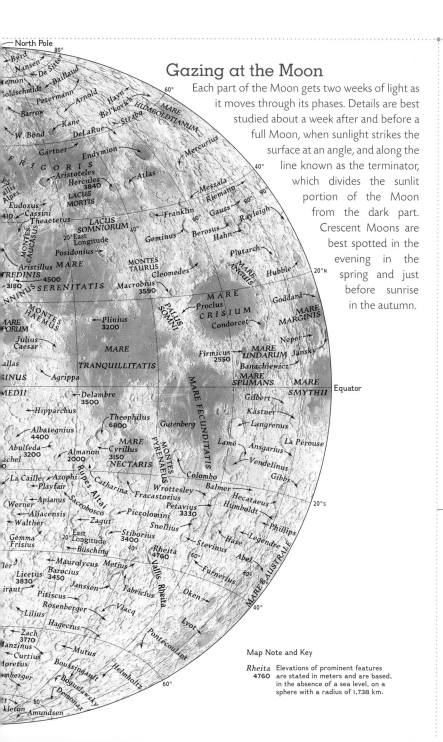

Gazing at the Moon

Each part of the Moon gets two weeks of light as it moves through its phases. Details are best studied about a week after and before a full Moon, when sunlight strikes the surface at an angle, and along the line known as the terminator, which divides the sunlit portion of the Moon from the dark part. Crescent Moons are best spotted in the evening in the spring and just before sunrise in the autumn.

Map Note and Key

Rheita
4760   Elevations of prominent features are stated in meters and are based, in the absence of a sea level, on a sphere with a radius of 1,738 km.

I N THE NIGHTTIME SKY you can see the stars, but if you look closely you can also see many of the objects humans have put into the sky over the past 50 years—spacecraft, satellites, and even space junk.

First in Orbit

The word "satellite" can mean any object that orbits a planet. Our Moon, for example, is Earth's natural satellite. Humankind's first experiment with launching objects into orbit occurred more than 50 years ago when, on October 4, 1957, the Soviet Union launched Sputnik I, the first man-made satellite; the first American satellite, Explorer I, would follow on January 31, 1958. In addition to launching the space age, Sputnik I also launched the space race, an intense competition between the U.S.S.R. and the U.S.A.

Manned Spacecraft

Almost as quickly as we launched unmanned satellites, we began to launch "manned" spacecraft as well. The first of these was Sputnik II, the second satellite launched into orbit, on November 3, 1957. It held a passenger named Laika, a small dog whose survival into orbit suggested that humans might survive launch as well (sadly, Laika died while in orbit, as a result of stress and overheating).

The first human to orbit the Earth was Soviet cosmonaut Yuri Gagarin, who was launched into space on the Vostok 1 mission on April 12, 1961. Gagarin made a single orbit of the Earth, but unlike Laika survived the experience, exiting his space capsule once it had reentered the atmosphere to parachute back down to the ground. When he landed, his first mission was to find a phone to call Moscow.

American Alan Shepard soon followed on May 5, 1961, with the Mercury-Redstone 3 mission, with John Glenn becoming the first American to orbit the planet in February 1962, in the Mercury-Atlas 6 mission. The most famous are the Apollo missions, which placed the first people on the Moon, beginning on July 20, 1969, with the Apollo 11 mission and continuing through December 19, 1972, with Apollo 17. To date, no one other than the Apollo astronauts has set foot on the Moon; in 2004, President George W. Bush announced his intention to return humans to the Moon by 2020.

The single most famous spacecraft is probably the space shuttle—or more accurately, the fleet of space shuttles used by the United States between 1981 and 2010 (when the remaining shuttle fleet is scheduled to be retired). Unlike previous manned spacecraft, the shuttles were designed to be reusable; they were launched from a rocket but were flown back to Earth like an airplane. The five shuttle orbiters—the *Columbia,* the *Challenger,* the *Atlantis,* the *Endeavour,* and the *Discovery*—have flown more than 120 missions between them, although not without loss: The *Challenger* was destroyed during a failed launch on January 28, 1986, and the *Columbia* disintegrated on reentry on February 1, 2003. In both cases the entire mission crew perished.

Space Stations

For people to stay in orbit for a long period of time requires a space station. The Soviet Union (and later Russia) has tried several of these, beginning with the Salyut 1 space station in 1971; there would be several others in this line, with Salyut 7 concluding its mission in 1986, to be followed by the Mir space station, which would operate from 1986 through 2000. The United States launched Skylab in 1973, which hosted three crews through 1974. Right now, there is a single space station in orbit: the International Space Station (ISS), a joint effort of the U.S.A., Russia, Japan, Canada, and 11 European countries. It received its first crew in November 2000 and is expected to remain in operation through 2016. Because of its size (it's the largest space station ever constructed), it is very easy to see in the night sky, making it a favorite among amateur sky watchers. Check out NASA's Orbital Tracking site *(spaceflight.nasa.gov/realdata/tracking)*, which has information on where the ISS is in its orbit at any time.

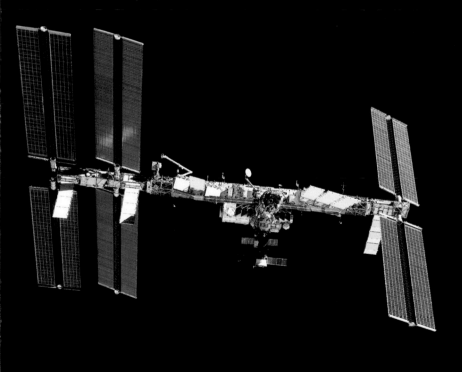

International Space Station

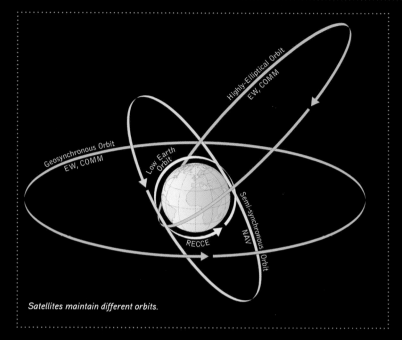

Highly-Elliptical Orbit
EW, COMM

Geosynchronous Orbit
EW, COMM

Low Earth Orbit

Semi-synchronous Orbit

RECCE

NAV

Satellites maintain different orbits.

SINCE SPUTNIK I's launch in 1957, humans have placed thousands of objects into near space, and many of these still orbit our planet; these are "artificial satellites," and what we usually refer to when we use the word "satellite."

Beginnings

The idea of launching a satellite into orbit isn't new; the concept was used in science fiction, most notably by Jules Verne, in the mid-19th century. The scientific argument for launching and using satellites began in 1903, when Russian scientist Konstantin Tsiolkovsky calculated the orbital speed required to launch an object into orbit around the Earth. In the 1920s, Slovene scientist Herman Potocnik further refined the concepts behind the geostationary satellite (that is, one whose position stays fixed over one spot on the planet), and in 1945, scientist and science fiction writer Arthur C. Clarke put forth the idea of using these sorts of satellites for high-speed, global communication.

Modern Satellites

Today, satellites are no longer the frontier of a space race; they are used largely for global communications and for scientific research and exploration. There are communications satellites, for routing phone and Internet communication as well as other data; weather satellites, to help track and predict weather systems across the globe; navigational satellites, whose

The first man-made satellite to orbit the Earth, Sputnik I

data power modern-day global positioning systems that let people, cars, and ships know where they are on the planet to within a few yards; and astronomical satellites, which collect data about the universe (the Hubble Space Telescope is a famous example of this kind). There are also reconnaissance satellites, used by the United States and others, whose uses are generally kept secret from the public.

Observation

Currently there are hundreds of operating satellites hovering over our planet, and many of them are observable by eye or through binoculars, because they are large enough—usually over 20 feet (6.1 meters) in length—and close enough—between 100 and 500 miles (161 and 805 kilometers) up—to be easily detected. A good time to view satellites is in the twilight, an hour or so after sunset or an hour or so before sunrise. This is because the satellite will reflect the light of the Sun, which is below the horizon. You can use the Internet to find where viewable satellites will be in the sky. One resource for this is Heavens-Above (http://heavens-above.com/).

Communications satellite

THE THOUSANDS of things that humans have launched into space over the years prompt an interesting question: What happens to all those things when we're done with them? In the case of the manned orbital missions, Moon landings, and shuttle missions, those spacecraft come back to Earth, through planned reentry into the planet's atmosphere. But with unmanned satellites, it can be a different story. Very often when we're done using a satellite, that satellite doesn't go anywhere—it just keeps orbiting the planet. For example, the Vanguard I satellite, the fourth satellite launched into orbit, has been circling the Earth since March 17, 1958, and to date has made nearly 200,000 transits around the globe. Its mission ended in 1964, but at this point it is likely to remain in orbit, just zooming around out there, for another 200 years.

Clutter & Debris

There is a term for nonfunctioning satellites like these: "space junk"—objects cluttering up orbital paths and the near space around the planet. But space junk is not limited to defunct satellites—everything we've sent up into space that remains in orbit eventually becomes space junk. This ranges from the rocket boosters that get satellites into space, down to paint flakes and explosion debris, and even actual garbage jettisoned from the Mir space station. And astronauts lose things from time to time: In 2006 shuttle astronaut Sunita Williams's digital camera accidentally broke free while she was performing an extravehicular activity; it's now floating out there in space. It joins another camera lost by astronaut Ed White on the first ever American space walk during the Gemini 10 mission. All told, there are an estimated 600,000 objects 0.4 inch (1 centimeter) wide or larger in orbit around the planet.

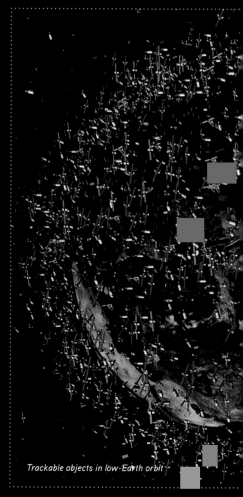

Trackable objects in low-Earth orbit

Hazardous Skies

Does it matter if this junk is floating out there in orbit? It does, because just as an object left on a road becomes a hazard to drivers, so does this junk become a hazard to functioning spacecraft and satellites. In 1991, the space shuttle *Discovery* had to maneuver to avoid debris from a defunct Russian

Space shuttle Discovery

satellite. In 2001, the entire International Space Station had to be nudged into a slightly higher orbit by the space shuttle *Endeavour* to avoid the risk of colliding with the spent upper stage of a Soviet SL-8 rocket, which had been circling the planet for three decades. Even a minor collision with space debris could send the ISS crew scrambling to abandon the station; a major collision would be catastrophic.

Another hazard of space junk is what happens to it when it falls out of orbit unplanned, raising the potential of hitting human habitation or spreading radioactivity. In 1979, the American space station Skylab fell from orbit earlier than expected, breaking up over the Indian Ocean and Australia, where fragments of the space station rained over the western part of the country. The Australian city of Esperance fined the U.S. government $400 for littering, but to date, the U.S. has not paid the fine.

Somewhat more ominously, in 1978 a Soviet satellite reentered the Earth's atmosphere over Canada, spewing radioactive debris into the Canadian Arctic. Canada charged the U.S.S.R. more than six million dollars for the cleanup, of which the Soviet government eventually paid half.

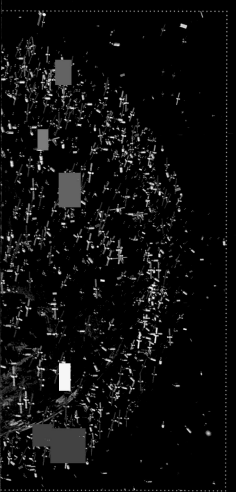

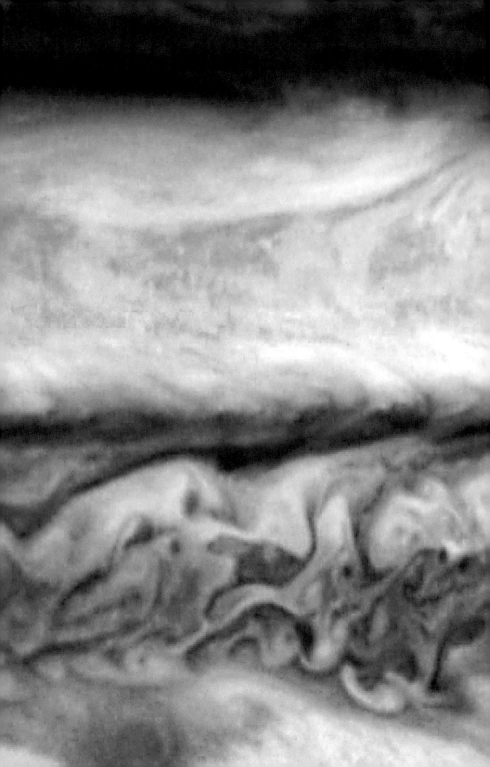

The Planets

5

Jupiter's vibrant atmosphere swirls and churns.

What Is a Planet?

TO THE ANCIENT Greeks and others they were the universe's vagrants, brilliant points of light that shared an unusual feature: They sauntered around a celestial orb whose occupants were surmised to occupy fixed positions. Sometimes these planets—the name derives from the Greek word for "wanderer"—

> **SKYFACT**
> When three celestial bodies array in a straight line, it is called opposition or conjunction. A more general term is syzygy.

would disappear from the sky altogether, then return to view at another time of year.

Planetary Qualifications

We now know that the visible changes in a planet's apparent position, relative to the stars, result from its orbit around the Sun. Recent debate over the status of Pluto, which is no longer considered a planet, also produced a new definition for what separates these wandering bodies from other objects in the solar system.

Planet Earth

The ability to peer deeper into space has led to the discovery of around 300 exoplanets—planets circling a star other than the Sun. It is a large enough number to raise doubts about whether our solar system will prove as singular as we typically regard it. Although the planetary systems discovered so far are different from ours—with gas giants and distended orbits the norm—astronomers estimate that as many as 10 percent of stars may prove to have planets.

to nonplanetary status; and created classifications for items like comets, asteroids, and dwarf planets. Had the Greeks known about Jupiter's dozens of moons or the icy detritus circling in the distant Kuiper belt, they may have surrendered their notions of the universe as a polished and perfectly ordered sphere.

Beginnings

Most of the objects in our solar system share an origin in the nebulous solar cloud that eventually coalesced into the Sun. As the cloud rotated, it developed a round proto-Sun surrounded by a wide, flattened disk of dusty gases. In the proto-Sun's heart pressure increased, triggering the nuclear fusion that set the Sun's life cycle into motion.

In the early 21st century, discoveries of planetlike objects in the disant Kuiper belt caused the the International Astronomical Union to set three conditions to qualify for official planetary status. First, planets are bodies that orbit the Sun. Second, they have a roughly round shape as a result of their own gravity. And third, they are large enough to have cleared their orbital path of debris. As a result, Pluto was demoted to dwarf planet status.

Yet that has not settled all the mystery surrounding Earth, its seven planetary companions, their satellites and rings, and the other residents of the solar system. Though it is estimated that as many as 10 percent of stars have at least one planet in orbit, neighborhoods as crowded as ours may prove to be rare.

Since the days when Babylonian, Greek, and Egyptian astronomers tracked bodies like Jupiter and Venus with the naked eye, scientists have added planets visible only through a telescope; changed the definition and downgraded one of those (Pluto)

The material left circling around the Sun, meanwhile, began to evolve on its own into separate bodies. Bits of rock and ice began sticking together and accumulating as planetesimals, gathering mass and eventually building up to planetary size. Clouds of gas were drawn into orbit around them. Closer to the Sun, the objects tended to be denser and rockier, with lighter atmospheres or none present at all. These formed what are now known as the terrestrial planets: Mercury, Venus, Earth, and Mars. Farther away, where the Sun's vaporizing effects were less profound, large clouds of gas collected around cores of rock and ice, forming the last four planets, what are now called the gas giants—Jupiter, Saturn, Uranus, and Neptune.

Discovery of the Planets

SOMETIME around 1600 B.C., under the reign of Babylonian ruler Ammisaduqa, astronomers began carrying out a systematic study of Venus. They assembled 21 years worth of data on the planet's motion, some of the earliest astronomical records in history. Across ancient cultures planets were worshipped and given godlike properties.

SKYFACT

Believing that another body was influencing Uranus, astronomers hunted for a ninth planet in the early 20th century. American astronomer Clyde William Tombaugh detected Pluto in 1930 as a small speck on photographs—too small to influence Uranus, and ultimately too small to remain a planet.

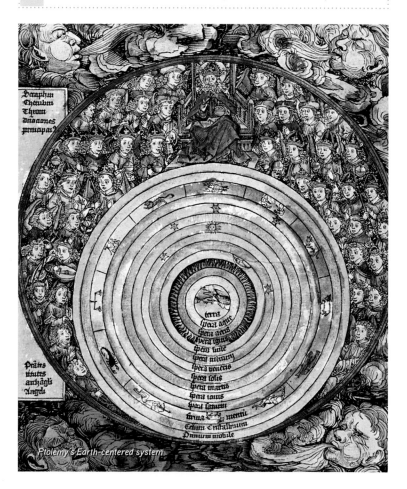

Ptolemy's Earth-centered system

Naked-Eye Planets

Early stargazers paid much attention to Venus and the four other naked-eye planets: Mercury, Mars, Jupiter, and Saturn. It is no surprise that they took on such importance. The five are among the brightest objects in the sky. Venus registers a dazzling -4 on the stellar magnitude scale, while Jupiter at its brightest is almost a -3.

Still, planet watching took a relative degree of sophistication. Egyptian astronomers, for example, referred to Mercury and Venus as "companions of the Sun" because they knew that those two are always spotted in the Sun's vicinity. A product of their place as the two planets closest to the Sun, Venus is only visible a few hours after sunset and before sunrise, while Mercury is visible at twilight or at dawn.

Enter the Telescope

The other planets in the solar system remained hidden until the invention of the telescope. English astronomer William Herschel is credited with discovering Uranus in 1781, though the planet had been spotted as early as 1690 and was frequently recorded as a star on celestial maps. Herschel felt that the object did not look particularly "starlike," and first wrote of it as a comet. After following its movement for a few weeks, he determined instead that it was a planet, the first to be discovered beyond Saturn.

Neptune, by contrast, was "felt" before it was seen. In the early 1800s, discrepancies turned up in Uranus's motion, leading to the conclusion that

William Herschel

John Cameron

something was tugging at it. French scientist Urbain-Jean-Joseph Leverrier and English astronomer John Couch Adams independently calculated the approximate position and size of the new planet. In 1846, a team at the Berlin Observatory found Neptune orbiting near where anticipated.

THE STORY OF Gods & Planets

The five planets visible to the unaided eye were named after figures from Roman mythology, a tradition followed when Uranus and Neptune were discovered. Mercury, visible only for short periods of time, was named after the fleet messenger god. Venus was named for the goddess of love, no doubt for its shining brilliance in the sky. Mars was named after the god of war. Jupiter was named after the Roman überdeity who ruled the heavens. Sluggish Saturn was named for the god of agriculture; Mesopotamian astronomers simply referred to it as "the old sheep."

The Solar System at a Glance

THE SOLAR SYSTEM is dominated—as the name implies—by the Sun. The Sun's gravity holds the structure in place, its energy fuels life on Earth, and it accounts for all but a minuscule portion (about 0.2 percent) of the solar system's mass. But that tiny fraction still includes a diverse array of celestial objects spread out over a distance of about 100,000 astronomical units—from the innermost planets to the outer reaches.

Comets, Asteroids, & Meteors

That massive distance is the approximate radius of the solar system, from the Sun's central point to the distant Oort cloud, home to billions of massive, drifting ice balls that, when dislodged, can speed through the solar system as fiery comets. The Oort cloud is one of three comet breeding grounds. The other ones lie much closer: in the Kuiper belt and its outer

SKYFACT

Need a way to remember the order of the planets in the solar system (Mercury, Venus, Earth, Mars, Jupiter, Saturn, Uranus, Neptune)? The International Astronomical Union suggests using this mnemonic phrase: "My very educated mother just served us nachos."

Backyard Guide to the Night Sky

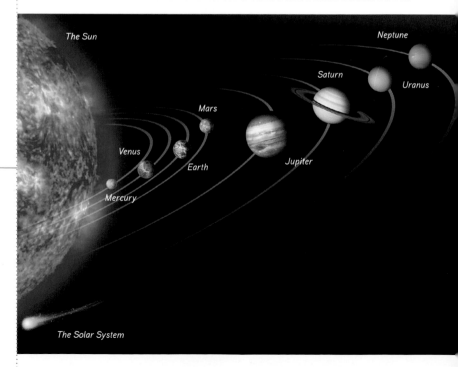

The Sun

Neptune

Saturn

Uranus

Mars

Venus

Earth

Jupiter

Mercury

The Solar System

SKY WATCHERS: Johannes Kepler

German mathematician and astronomer Johannes Kepler offered a bridge between the lingering classical notions of a solar system built from perfectly ordered spheres and circular motions and the more chaotic reality that became apparent as the telescope and subsequent inventions allowed better observation. Studying in the late 16th and early 17th centuries, Kepler set out to prove that the underlying motions of the planets did follow a pattern—even though it had become apparent that they were not moving in the perfect circles assumed by earlier astronomers. He ultimately formulated three laws that describe the elliptical motion of celestial bodies around each other—rules that explained what was happening with the planets, but that also apply to moons, comets, and other objects. Even as the cosmos became more crowded, Kepler's laws of planetary motion showed that an underlying order was still maintained.

extension, the scattered disk. The Kuiper belt and scattered disk are regions of icy and rocky debris, as well as near-planet-size objects, that begin just beyond Neptune's orbit and extend to about 1,000 AUs.

Closer in, a 340-million-mile (547-million-kilometer) gap between Mars and Jupiter is littered with asteroids, hunks of rock from the solar system's early days. This "asteroid belt" marks the border between the terrestrial planets and the gas giants. Its existence is attributed to Jupiter's massive gravitational field, which prevented any other planet from coalescing in that space and left the scattered bits of debris to collide and pulverize each other into smaller and smaller pieces.

The brightest asteroids can be spotted with a backyard telescope. More often, however, they are noticed when smaller ones come burning through Earth's atmosphere as meteors, a catchall term for falling interstellar debris, more commonly known as shooting stars. Those that make it to Earth are known as meteorites.

The Planets

The most easily located objects of the solar system (excluding our own Earth, the Sun, and the Moon, of course) are the planets. They are spread across a distance ranging from Mercury, at roughly 36 million miles (58 million kilometers) from the Sun, to Neptune, whose orbit keeps it at an average distance of around 2.8 billion miles (4.5 billion kilometers) from the central star. Because of the dynamics of the solar system's formation, they travel close to the ecliptic, along a plane laid out by the spinning motion of the gas cloud that formed the Sun. And following along with them—particularly giants like Jupiter—are dozens of moons, many of which are good telescope targets, and some of which can be spotted with binoculars.

Viewing the Planets

OBSERVING THE PLANETS can involve a range of effort and equipment, depending on the target and the level of detail that is desired. Venus—the brightest object in the sky after the Sun and Moon—is a compelling sight with the

naked eye, as is Jupiter. But while Saturn can be seen unaided, spotting its popular numerous rings requires a telescope with an aperture of more than 3 inches; Jupiter's moons can be seen as starlike points through binoculars, while a telescope with at least 2.5 inches of aperture will display the planetary disk flanked by its satellites. Finding distant Neptune will require both the use of a chart or software and practice to discern it from the neighboring stars.

Venus and Jupiter are two of the brightest planets.

Resources

Charts, almanacs, and other resources are helpful for determining the best viewing times. Published resources will typically place the planets month to month in different constellations, mostly along the zodiac. Special events will also be featured in places like *Sky & Telescope* magazine's weekly summary of the top objects in the sky and *Astronomy* magazine's online "Star-Dome." Space.com also has monthly summaries of planetary locations.

THE STORY OF Planet Vulcan

Neptune's discovery began the hunt for another planet that was thought to be causing changes in Mercury's orbit. Urbain-Jean-Joseph Leverrier, one of the astronomers who had foreseen Neptune's presence, calculated that a planet about the size of Mercury and orbiting half as close to the Sun might affect its orbit. Leverrier in 1859 announced that he had discovered a new world, Vulcan, speeding around the Sun every 20 days. What had been observed, however, proved to be a sunspot. It took Einstein's theory of general relativity, developed early in the 20th century, to explain the slippage in Mercury's orbit.

Rules of Thumb

For anyone hoping to spot the planets without gear or charts, however, some rules of thumb will help. For example, Mercury, the closest to the Sun, is visible only right at dusk and dawn. Venus is best viewed when the planet forms a roughly 45-degree angle with the Sun and Earth. Being much brighter, Venus also gets higher in the sky than Mercury. The interior planets disappear twice in their orbital year—once when they pass between the Earth and the Sun (a point called inferior conjunction), and once when they pass into superior conjunction at a point that is, from Earth's perspective, behind the Sun.

The exterior planets, from Mars and beyond, can appear anywhere along the ecliptic. The best viewing is typically when the Earth passes between the planet and the Sun, a situation called opposition that is akin to a full Moon. It happens roughly once a year. The exception is Mars, whose orbital period is much closer to Earth's and thus makes it harder for us to catch up.

Mercury

RADIUS: 1,516.0 mi (2,439.7 km)
MASS: 3.3022 x 10^{23} kg
DISTANCE FROM SUN: 35,983,095 mi
(57,909,175 km)

ORBITAL PERIOD: 88 Earth days
ROTATIONAL PERIOD: 58.6 Earth days
NUMBER OF SATELLITES: none
MAGNITUDE IN SKY: up to -1.9

THE SMALLEST and inner-most of the planets, Mercury, though close enough to be seen with the naked eye, is also one of the more difficult to spot. It lies less than half the distance between the Earth and the Sun—about 0.39 AU, or 36 million miles (58 million kilometers)—and circles the star every 88 days. It is, there-fore, frequently hidden from view

SKYFACT

Only one space mission—Mariner 10—has visited Mercury as of 2007. But NASA's MESSENGER (MErcury Surface, Space ENvironment, GEochemis-try, and Ranging) probe, having left Venus, did its first flyby of Mercury in January 2008 and is scheduled to slip into orbit in 2011 for a year of study.

Mercury

in the glare of sunlight. During a typical year, it is visible six times, for a couple of weeks, either low in the west after sunset or low in the east before sunrise. Almanacs, astronomical software, or astronomy magazines contain tips on how best to locate it.

On the Planet

Little was known about the surface of the planet until the mid-1970s, when photographs from the Mariner 10 spacecraft showed it to be barren, rocky, and strewn with craters. It has, in essence, no atmosphere, and is a land of temperature extremes: The region around the equator may range from 800°F (427°C) during its day to -300°F (-184°C) during its long nights. Its molten iron core makes up about 65 percent of the planet.

Viewing Mercury

The best times to hunt for Mercury are during the fall and spring, when the ecliptic stands at its sharpest angle relative to the horizon, and the tiny planet reaches its widest separation from the Sun. For Northern Hemisphere observers, Mercury's evening appearances in March and April, or its morning apparitions in September and October, are the prime times.

If the atmosphere cooperates and the horizon is clear and unobstructed, the planet might be seen on consecutive nights for as long as three weeks running before it disappears into one of its frequent conjunctions. But turbulence is more likely to occur near the horizon, so the image may prove little

Leverrier observed Mercury's odd orbit.

better than a shaky speck in your eyepiece. Similar to Venus, it will appear to go through phases—gibbous, half-lit, then crescent—as it moves around the Sun and catches up with Earth. Binoculars are useful for locating Mercury in a twilit sky but, of course, don't show its phases, as a telescope can.

THE STORY OF Einstein & Mercury's Orbit

A steady movement forward in the location of Mercury's closest approach to the Sun—its perihelion—was a mystery. Albert Einstein solved it when he published his general theory of relativity in 1915, using it to explain the eccentricities in Mercury's path. Einstein explained gravity not as attraction between bodies, but as a warp in space. In Mercury's case, the Sun's massive gravitational field was contorting the fabric of the universe and causing Mercury's orbit to change and slip ahead.

101

Venus

RADIUS: 3,760.4 mi (6,051.8 km)
MASS: 4.8685 x 10²⁴ kg
DISTANCE FROM SUN: 67,237,910 mi
(108,208,930 km)

ORBITAL PERIOD: 224.7 Earth days
ROTATIONAL PERIOD: 243 Earth days
NUMBER OF SATELLITES: none
MAGNITUDE IN SKY: up to -4.5

WITH A DIAMETER and mass similar to Earth's, and a 225-day orbit around the Sun, Venus until the 1950s was thought of as sort of a sister planet—perhaps with a lush, tropical climate waiting to be discovered. The planet named for the Roman goddess of love, however, has hardly proved welcoming.

On the Planet

Much hotter than tropical, Venus is the hottest planet in the solar system, with a surface temperature in excess of 860°F (460°C). Though farther from the Sun than Mercury, Venus has a surface subject to a runaway greenhouse effect: Thick with carbon dioxide, the planet's dense atmosphere traps the Sun's heat.

Venus is also blanketed in clouds of sulfuric acid, a highly reflective covering that makes it the brightest of all the planets and stars but that hides its surface from view. Details about surface conditions have all come through space probes, beginning with the U.S. Mariner 2 in 1962 and continuing to the present. NASA's Magellan orbiter mapped the Venusian surface by radar between 1990 and 1994.

Viewing Venus

As one of the two interior planets, Venus will always be seen in the hours after sunset and before sunrise; it never moves too far from the Sun. But its proximity to Earth and its similar orbital year also mean that it is frequently visible for at least some part of the night.

At first it will appear low to the horizon at nightfall in the western sky, the point in its course when the planet is at its smallest and dimmest. As it continues on its path, Venus gradually moves farther from the Sun—from an

Venus

SKYFACT

Galileo's observations of Venus helped validate the heliocentric model of the solar system. Because Venus never grew beyond its gibbous state, he deduced that the planet was ducking behind the Sun between one gibbous phase and the next. Were it circling the Earth, its phases would resemble the Moon's.

earthbound perspective—and higher in the sky. Reaching its greatest eastern elongation (the greatest angular separation from the Sun), the planet will be at its highest point in the sky, have more than doubled in apparent size, and reached it brightest magnitude of -4.5.

As Venus gradually overtakes Earth from behind, it grows still larger and brighter. To the unaided eye, that will mask the fact that Venus moves through phases similar to those of the Moon. At its two elongations, it is half lit. Approaching inferior conjunction, roughly every 18 months, it enters a crescent phase that can be discerned through binoculars (and, for a lucky few with excellent vision, by the naked eye), before disappearing as it passes between Earth and the Sun.

After overtaking Earth in its orbit, Venus now becomes the morning star, rising just before the Sun, at first, but appearing a little earlier and higher each evening as it speeds away to its greatest western elongation. Eventually it disappears into superior con-

Bright Venus alongside the Moon

junction, on the other side of the Sun, before a new apparition begins.

Along with tracking Venus's phases, observers look for faint changes in the cloud covering, or try to spot the "ashen light" seen sometimes on the unlit portion of the planet. Blue or violet filters can help bring out cloud details. Viewing in twilight is best, as the planet's brightness can make it overpowering in a dark sky through a telescope. For more detailed viewing, consider observing the planet before the Sun fully sets or as it is rising.

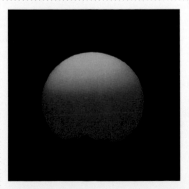

THE STORY OF The Transit of Venus

Venus's orbit occasionally takes it across the face of the Sun at inferior conjunction. When these transits occur, Venus appears as a dark spot on the Sun's surface. In the late 1760s an anticipated transit inspired England's Royal Society to send a team to observe it from the southern latitudes. Naval authorities appointed James Cook to captain the H.M.S. *Endeavour*. Along with the astronomical mission, he was tasked with finding a southern landmass that was assumed to exist, which led him to chart New Zealand and parts of Australia.

Mars

RADIUS: 2,111 mi (3,396.2 km)
MASS: 6.4185 x 10²³ kg
DISTANCE FROM SUN: 141,633,260 mi (227,936,640 km)

ORBITAL PERIOD: 687 Earth days
ROTATIONAL PERIOD: 24.6 Earth hours
NUMBER OF SATELLITES: 2
MAGNITUDE IN SKY: +1.8 to -2.91

IN MANY WAYS, Mars is Earth's closest cousin. It is tilted similarly to the ecliptic plane, and thus experiences seasons and weather. Although Mars's rotation axis tips about as much as Earth's, it points in a different direction. As a result, whenever Mars comes to opposition for terrestrial observers, the time of year on Mars is always one season ahead of that on Earth.

Viewing Mars

Mars can be seen with the naked eye, but its apparent size and brightness vary markedly during its 687-day orbit. Binoculars will show its reddish color, but a telescope of modest size—

Mars

Jupiter

Mars

Saturn *Venus*

Mars moves along the ecliptic with Jupiter, Saturn, and Venus.

a 5-inch refractor or 8-inch reflector—will reveal the planet's details. When the air is clear, darkish lava and boulder fields can be seen, as well as the frozen polar caps.

Mars's best views are when it is in opposition. Because its orbit more closely matches Earth's, those moments are infrequent compared to the roughly annual oppositions of the gas giants. From one opposition of Mars to the next takes roughly 780 days. The red planet's eccentric, stretched-out path brings it as close as 35 million miles (56 million kilometers) to Earth roughly every 17 years, when Mars is both at perihelion (closest point to the Sun) and at opposition for Earthlings.

Mars begins each new apparition near the eastern horizon, rising before sunrise and moving progressively farther from the Sun and higher in the sky.

As Earth overtakes it in orbit, Mars's motion appears to stop before it begins a roughly three-month journey westward. After coming into opposition, it will resume an eastward course, appearing ever lower in the evening sky before slipping into the Sun's glare.

THE STORY OF A Martian Invasion

Mars occupies a special place in the human imagination as a planet that could support and launch an invasion of Earth. In 1938, dramatist and filmmaker Orson Welles adapted the novel *The War of the Worlds* into a radio play that was presented—on the day before Halloween—as a modern-day newscast featuring on-the-scene reports from the New Jersey scene of a Martian attack. The broadcast prompted a flood of worried calls to local police and front-page headlines in the next day's paper noting—for those who missed Welles's explanation at the end—that the broadcast was a work of fiction.

Mars: *Up Close*

A MANNED TRIP to Mars remains a goal of many national space agencies around the world. The danger and logistics of such a mission—with round-trips of as much as 900 days for the astronauts involved—make it a distant reality, still most likely decades away. But with more than 30 unmanned missions launched or attempted by the United States and other countries since the 1960s, Mars already has been one of the more frequently visited planets.

Mars's dramatic topography

Visitors to Mars

The first flyby probe, Mariner 4 in 1965, revealed a barren and rocky landscape, with a dusty reddish surface made mostly of iron oxide, silicon, and sulfur. The atmosphere is thin—less than one percent of Earth's and made mostly of carbon dioxide—and cold, with temperatures averaging minus 85°F (-65°C). Cratered highlands dominate the southern hemisphere and smoother plains are in the north.

With a planetwide desert and, in some areas, windswept dunes, Mars also has some of the most dramatic topography in the solar system, including an equatorial region called Tharsis that is home to several immense volcanoes. The area appears to have been created over billions of years by successive volcanic eruptions. Also located near the equator, the impressive Valles Marineris canyon network spans about 2,400 miles (3,862 kilo-

SKYFACT

The solar system's biggest volcano—over 13 miles (21 kilometers) high and 300 miles (483 kilometers) wide—is found on Mars.

A Mars rover, Spirit, captures a sweeping planetscape.

American astronomer Percival Lowell is known for one of modern astronomy's most imaginative miscalculations. He was a chief proponent of the idea that faint linear markings on Mars's surface were evidence of an irrigation system built by an advanced civilization, since deceased. Though the idea was never widely accepted (and finally laid to rest when Mariner 4 sent back the first photos of the barren planet in 1965), Lowell publicized his theories in a book, *Mars and Its Canals*. Why canals? The theory may have started with an overly aggressive translation of work done by Italian astronomer Giovanni Schiaparelli, who in 1877 spotted latticelike lines he referred to as *canali*, by which he meant natural channels in the landscape, not man-made structures.

meters) and in spots is more than 29,000 feet (8,839 meters) deep.

Rovers & Surveyors

The most recent missions to Mars have returned some of the most intriguing information and seem to support the notion that the planet was once covered with lakes, seas, and other bodies of water. As the two robotic rovers Spirit and Opportunity trekked over the planet in 2004, some of the geologic samples showed evidence of having been submerged. Visual evidence from the Mars Global Surveyor, meanwhile, showed what appeared to be shorelines, gullies, and old riverbeds. One theory is that Mars went through a "wet period" perhaps as much as four billion years ago. With water locked in its ice caps and apparently in layers of permafrost around the chilly planet, it is thought that shifts in the planet's orbit and the degree of tilt on its axis changed it from a warmer, wetter world.

107

Asteroids & Dwarfs

BETWEEN MARS and Jupiter, the territory between the terrestrial planets and the gas giants is populated by millions of rocky objects. Because of the influence of massive Jupiter, they never coalesced into a planet. They are called minor planets or asteroids.

The Asteroid Belt

Of the profusion of bodies in the asteroid belt, more than 120,000 have been cataloged and 13,000 have been named—an honor roll of girlfriends, religions, and favorite writers, among other things. Each member of The Beatles has one named after him.

There may be several hundred thousand smaller asteroids out there. Though they are too small to be spotted, their presence can be felt in other ways. Objects in the asteroid belt frequently collide, and flying debris from time to time comes burning through Earth's atmosphere as a meteor. Occasionally more durable pieces reach Earth's surface as a meteorite.

Two smaller fields have collected along Jupiter's orbital path, the so-called Trojan asteroids. A few thousand

> **SKYFACT**
> Before being named a dwarf planet, Ceres was by far the largest asteroid—the next in size, Pallas and Vesta, are less than half as big. While tens of thousands of the largest asteroids have been found, it is conjectured that there could be as many as a million with a diameter of at least a half mile.

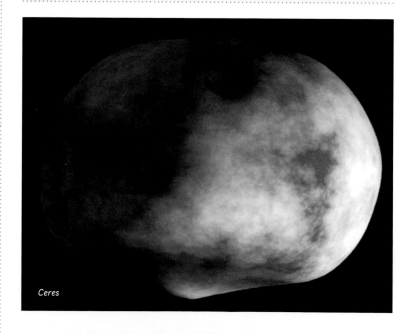

Ceres

Asteroid impacts can leave giant craters.

other near-Earth asteroids roam the region near our planet. One among them stands out. With a diameter of 580 miles (933 kilometers), Ceres is no longer considered an asteroid, but was elevated to dwarf planet status after the International Astronomical Union created that category in 2006 to define planetlike objects that orbit in heavy company.

Seeing Asteroids

Hundreds of asteroids are theoretically within the range of even a modest 3-inch telescope. Finding them is another matter. Astronomical publications will provide special lists, known as ephemerides, for the most prominent ones. But remember, these objects will be faint and largely indistinguishable from the stars. There are two strategies for picking them out once the right area of the sky is located. Using a low-powered eyepiece, you can map (or photograph) what you see, and compare that with observations later in the evening, or on a subsequent night. to determine what has moved. Alternatively, you can compare the field of view with a sky chart: The asteroid will be the starlike object not on the map.

SKY WATCHERS:
Giuseppi Piazzi

Astronomers in the late 1700s were hunting for a planet in the space between Mars and Jupiter. It was a monk in Palermo, Italy, who found what everyone was looking for—or so everyone thought. Giuseppi Piazzi in 1801, observing at the Palermo Observatory in Sicily, located an object that was acting "planetlike." It was dubbed Ceres. Eventually, because of doubts about its size, shape, and presence among so many similar bodies, Ceres became simply the largest asteroid—until it was redesignated in 2006 as one of the new dwarf planets.

Jupiter

RADIUS: 44,423 mi (71,492 km)
MASS: 1.8987 x 10^{27} kg
DISTANCE FROM SUN: 483,682,810 mi (778,412,020 km)

ORBITAL PERIOD: 11.9 Earth years
ROTATIONAL PERIOD: 9.9 Earth hours
NUMBER OF SATELLITES: 63 moons
MAGNITUDE IN SKY: -1.6 to -2.6

JUPITER IS roughly 15 times as far way from Earth as is Venus, but at its most brilliant, it looms nearly as large and bright in the sky. Fortunately for sky watchers, Jupiter's 12-year orbit allows it to spend around a year in each constellation of the zodiac. It makes Jupiter an easy target beyond its peak magnitude.

Jupiter governs its own miniature solar system of 63 moons. The four largest ones—Io, Europa, Ganymede, and Callisto—are visibile with binoculars. First spotted

SKYFACT

Massive in most respects, Jupiter falls short on one front. Like Saturn and Uranus, Jupiter has a ring, first revealed by the Voyager 1 probe. But it is small and thin compared with the rings around the other planets, only about 3,700 miles (6,000 kilometers) wide—a puny crown for the planetary king.

Jupiter

by Galileo in 1610, they are referred to as the Galilean moons.

On the Planet

All told, its size, orbit, and constantly changing atmosphere make Jupiter one of the most popular planetary targets. Binoculars reveal at least four moons; in small telescopes details of Jupiter's disk begin to emerge, and with 6-inch or larger scopes these become easy to see.

With Jupiter we never see a solid surface, just the top of a roiling atmosphere that's thousands of miles deep. Currents swirling within this material create a large magnetic field, intense emissions of radio waves, and regular bursts of radiation. Descending through Jupiter's atmosphere, intense pressure creates a zone of metallic hydrogen that surrounds the planet's rocky iron core, which is molten and invisible to observers.

Viewing Jupiter

The real show for an astronomer is Jupiter's atmosphere. Jupiter's quick ten-hour rotation gives it a sort of lava lamp appearance, as swirling wind creates bands of faint pastel colors in the planet's thick mix of hydrogen, helium, methane, and ammonia.

Persistent storm systems add to the spectacle, the most famous of which is the Great Red Spot, a high-pressure zone in Jupiter's southern hemisphere that is about two times as big as Earth. The Great Red Spot—a circlelike patch below the equator—ebbs and flows but is typically visible,

Jupiter

Jupiter alongside the Milky Way

as are other white-hued storm systems. Even when the spot disappears, its absence can be detected in the Red Spot Hollow, a bend in the adjoining cloud belt. With your telescope, a light blue filter will make the lines between the atmospheric bands more definite, while yellow and orange filters will help bring out other details.

THE SCIENCE OF Life on Europa?

The Galilean moons were a focus of the Galileo orbiter launched from the space shuttle *Atlantis* in 1989. Collected data about the moons' conditions has begun to raise questions about the possibility of life on Europa. Its surface is a shell of ice, roughly 1 to 10 miles (1.6 to 16 kilometers) thick, covering a sea of briny water perhaps 60 miles (97 kilometers) deep. Tidal tugs from Jupiter and neighboring moons Io and Ganymede keep Europa's interior ocean warm, and there's an atmosphere, very thin but with molecular oxygen sputtered from the icy surface.

Saturn

RADIUS: **37,449 mi (60,268 km)**
MASS: **5.6851 x 10²⁶ kg**
DISTANCE FROM SUN: **885,904,700 mi (1,426,725,400 km)**

ORBITAL PERIOD: **29.4 Earth years**
ROTATIONAL PERIOD: **10.7 Earth hours**
NUMBER OF SATELLITES: **60 confirmed**
MAGNITUDE IN SKY: **0.6 to 1.5**

THE MOST DISTANT of the naked-eye planets, Saturn orbits nearly twice as far from the Sun as Jupiter, creaking along in a 29.5-year orbit that led astronomers in ancient Mesopotamia to dub it the "old sheep" of the sky.

But now Saturn is one of the solar system's standouts. The planet's razor-thin ice particle rings—170,000 miles (274,000 kilometers) side to side but in spots only a few dozen feet thick—are one of the first sights an amateur astronomer is likely to hunt.

SKYFACT

The ice particles that comprise Saturn's rings are as fine as the crystalline mist felt in a chilly fog and as large as a house.

On the Planet

To an observer, Saturn's atmosphere is all that you can see. The thick soup of gases—mostly hydrogen and helium—surround a body of liquid hydrogen

Saturn in natural and infrared (top) light

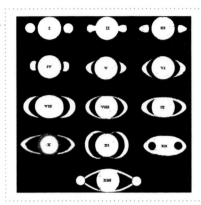

THE SCIENCE OF Saturn's Ears

When Galileo first viewed Saturn in 1610, he saw bulges at the planet's side that looked like moons. Two years later they had disappeared. Four years later, the bulges were back as large ellipses on either side of the planet. The changing observations were sketched in Galileo's notebooks, but he never solved the mystery. Some 40 years later, Christiaan Huygens explained it: A ring around Saturn, tilted to the ecliptic, would change perspective and appearance through its orbit—as the angle of observation was altered.

and a core of rock and ice. Its cloud top is covered in a chilly fog of ice crystals. The shearing forces that create such distinct and colorful cloud bands on Jupiter are less apparent on Saturn, which presents a more subdued image in shades of buff and white.

Viewing Saturn

Almanacs and sky charts will provide information about Saturn's location and rising and setting times. It will linger within a constellation for more than two years at a time. For a good view of the planet, a telescope is a necessity. Saturn at opposition is less than half of Jupiter's apparent size in the sky—about 20 arc seconds, compared with 50 arc seconds for Jupiter.

Saturn's rings cannot be distinguished through binoculars—a frustrating fact. But a small telescope will reveal them, while 6 inches of aperture will bring three moons into view, as well as details on the body of the planet. Keep in mind that the appearance of the rings does change as Earth and Saturn orbit the Sun. Saturn is tilted to the ecliptic—like Earth—and when the tilt is toward us we see a broad "top-down" view of the rings. The other side becomes visible when the planet is tilted away from us. Roughly every 14 years, however, Saturn's rings are tilted edgewise toward Earth, making them all but disappear from view—an event that will next occur in 2009.

113

CHAPTER 5: The Planets

Saturn: *Its Rings & Moons*

BEGINNING WITH Pioneer 11 in 1979 and Voyagers 1 and 2 in the 1980s, up close studies of Saturn have revealed much. The ongoing (as of 2008) Cassini-Huygens orbiter mission promises to teach us more.

Gaps in the Rings

Dutch astronomer Christiaan Huygens in 1659 first deduced that there was a ring around the planet. In 1675, Giovanni Cassini detected a gap in the ring, then thought to be a single, solid plane. The 2,600-mile (4,200-kilometer) gap known as the Cassini Division now distinguishes Saturn's outermost, 9,000-mile-wide (14,500 kilometers) A ring from the broad, bright, 16,000-mile-wide (26,000 kilometers) B ring. C ring is the faintest of the three, and about 10,500 miles (17,000 kilometers) wide. Édouard Roche, in the

SKYFACT

Jupiter has its Great Red Spot, and Saturn has a Great White Spot. Spots like these appear on Saturn's surface roughly every 30 years around the time of the northern hemisphere's summer solstice. While not fully explained, they may be large bubbles of ammonia released from the planet's interior.

Saturn's complex ring system

Rubble, dust, and ice compose Saturn's rings.

mid-1800s, conjectured that the rings were made of debris, either ripped apart by Saturn's gravity or prevented from reforming—stepping over what is now called the Roche limit.

Further study revealed more complexity. Along with the three major rings, more diffuse rings—D, E, F, and G—have been detected as part of the main ring structure. The continent-size gap discovered by Cassini, in fact, is simply a region where the collection of ice particles is less dense. The Voyager encounters showed that the rings were composed of thousands of smaller ringlets with a proliferation of ice particles orbiting in procession.

Cassini actually zipped through the ring plane (between the F and G rings) on arrival at Saturn in 2004 and found that the Cassini Division contained relatively more dirt than ice. The material

appeared similar to what was detected on the surface of Phoebe, one of Saturn's moons, lending credence to the idea that the rings are the pulverized remains of erstwhile moons or comets and asteroids that strayed too close.

THE SCIENCE OF Shepherd Moons

The structure of Saturn's rings has provided a massive laboratory for studying the intricacies of gravity. For example, scientists view some of Saturn's moons as critical in keeping order among the millions of ice particles that the rings comprise. The so-called shepherd moons—Pan, Atlas, Pandora, and Prometheus—straddle two of the rings in pairs and act like herders to keep them intact. Pandora and Prometheus maintain the thin and far-out F ring. Atlas orbits the outer edge of the A ring, while its partner Pan traverses and keeps open an inside area of the A ring known as the Encke gap.

Uranus & Neptune

Uranus

RADIUS: 15,882 mi (25,559 km)
MASS: 8.6849 x 10²⁵ kg
DISTANCE FROM SUN: 1,783,939,400 mi
(2,870,972,200 km)

ORBITAL PERIOD: 84.02 Earth years
ROTATIONAL PERIOD: 17.24 Earth hours
NUMBER OF SATELLITES: 27 moons
MAGNITUDE IN SKY: 5.3

FOR ANYONE training their telescope on these two most distant planets, be prepared for a less than dramatic outcome. Both are so distant that they will be relatively dim disks that will reflect the planets' dominant color but provide little detail. Locating them will provide practice in reading astronomical charts—they'll be necessary—and using your equipment to find a specific object.

Discovery

Astronomers of antiquity did not know about either planet. Both Uranus and Neptune were discovered and named in the telescopic age. Galileo came close in the 1600s. He actually laid

eyes on Neptune, but regarded it as a star because his equipment was not powerful enough to discern the distant planet as a disk. And despite his intense exploration of Jupiter in 1610, his telescope's narrow, 17 arc minute field of view meant that Uranus—hovering just two degrees from the giant planet at the time—went unnoticed. Even since discovery, their distance has remained an impediment to study. Though they were visited by the Voyager 2 mission, they have not drawn the attention given to the other giants: They are more than twice as far away.

Tilted Planet

Uranus lies almost on its side, tilted 98 degrees to its orbital plane, the possible product of a massive collision. It was spotted several times before William Herschel made the definitive observations in 1781; earlier astronomers recorded it as a star. It is technically at the outer limit for the unaided eye, but as a practical matter, binoculars will be needed. When spotted, the planet is a small blue-green disk. The

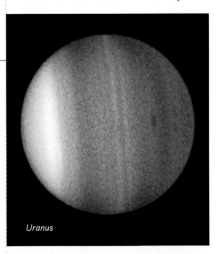
Uranus

Backyard Guide to the Night Sky

SKYFACT
Neptune's moon, Triton, is the only satellite with a retrograde orbit, moving opposite the planet's direction of travel.

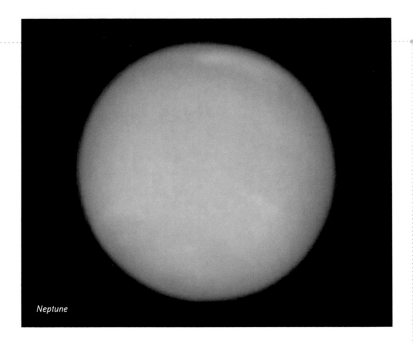
Neptune

Neptune

RADIUS: 15,387 mi (24,764 km)
MASS: 1.0244 x 10^{26} kg
DISTANCE FROM SUN: 2,795,084,800 mi
(4,498,252,900 km)

ORBITAL PERIOD: 64.79 Earth years
ROTATIONAL PERIOD: 16.11 Earth hours
NUMBER OF SATELLITES: 13 moons
MAGNITUDE IN SKY: 7.8

color is a product of methane gas in its upper atmosphere. Uranus's 5 major moons and more than 20 small ones are not visible to the unaided eye. Likewise, Uranus' network of 11 rings remain inaccessible to even the backyard telescope user.

Seeing Blue

Neptune was named after the Roman god of the sea and has maintained its distinctive blue color even under the closer observation of Voyager 2. It is now thought to possess a massive ocean around a core of rock, all beneath an atmosphere thick with hydrogen and helium. The bluish tint, however, comes from clouds of icy methane that blanket the planet.

Voyager 2 detected faint rings around Neptune, uncovered a storm system dubbed the Great Dark Spot, and brought the number of known moons to more than a dozen. This outermost planet will require a finder chart and some patience to distinguish it from the stars surrounding it in a telescope eyepiece. It has a gray-green disk about two-thirds the size of Uranus. Spotting the large moon Triton would require a telescope with at least an 8-inch aperture.

Pluto & Beyond

PLUTO

RADIUS: 715 mi (1,151 km)
MASS: 1.3×10^{22} kg
DISTANCE FROM SUN: 3,670,050,000 mi
(5,906,380,000 km)

ORBITAL PERIOD: 247.92 Earth years
ROTATIONAL PERIOD: 6.487 Earth days
NUMBER OF SATELLITES: 3 moons
MAGNITUDE IN SKY: 13.6

HOW ARE PLANETS distinguished from the other objects orbiting the Sun? Controversy over Pluto, a planet until 2006, helped crystallize the question beginning in the early 1990s.

Pluto's Status

Pluto's distant part of the solar system contains hundreds of objects left over from the solar system's formation. In 2003, an object—even larger than Pluto and with its own moon—was discovered, forcing a decision that changed the status of Pluto and objects like it. In debating Pluto's fate, the International Astronomical Union cited three tests for planetary status: The body must orbit the Sun, have a close-to-round shape created by

Viewing the solar system from Pluto

Looking at the Sun from Sedna, a dwarf-planet candidate

The list of dwarf planets is likely to grow in the coming years, perhaps by the hundreds. As of October 2008, the International Astronomical Union had acknowledged five. Pluto, Eris, Makemake, and Haumea are all plutoids that circle beyond Neptune. The fifth, Ceres, is located in the asteroid belt.

its own gravity, and have "cleared the neighborhood" of debris along its orbit. Pluto failed the third test, circling as it does in the company of many other "transneptunian" objects.

As a result, Pluto became the first dwarf planet, a new category. The 2003 discovery is the second dwarf, now known as Eris. The two were also placed in a special category of dwarf planets known as plutoids because they orbit beyond Neptune—a name that honors Pluto's former status.

Both reside in the Kuiper belt, a debris field that extends from about 35 AU to about 55 AU. Outside that lies a much wider scattered disk of small, icy bodies. The Kuiper belt is an area where planet forming appears to

have stopped. Objects are orbiting too slowly and are too spread out for them to bunch and collide and coalesce into larger bodies. They remain slow-circling planetesimals, protocomets that occasionally dislodge and travel through the solar system.

The Outer Reaches

Spotting Pluto requires at least an 8-inch telescope, and enough patience to watch for its movement over several nights against the stars around it in the field of view. Even sky charts will only provide a starting point. Pluto is so small and distant—a 1,400-mile-diameter (2,300 kilometers) object nearly 39 AUs beyond Earth—it will appear no different from the stars.

EVERYTHING YOU can see in the night sky—and many objects too faint to see—can be photographed. While astrophotography can employ complex computer-guided telescopes and specialized cameras, some of the most compelling images of the night sky require no more than the "off the shelf" camera gear.

Sky Subjects

The sky presents an inexhaustible array of subjects for a photographer. Landscapes that look conventional by day take on an exotic character when photographed by night using nothing but moonlight and starlight for illumination. Add a colorful display of auroras to the sky and you have a unique award-winning image. Wait for nights when the Moon lies close to bright planets and you have the makings of an eye-catching composition framing earth and sky. Many nightscapes can be captured with exposures no longer than 30 seconds. But extend the exposure time to minutes or an hour, and you get a sky filled with the colorful trails left by stars as they wheel about the heavens.

The vibrant colors of an aurora, a popular subject

Placing the camera on a telescope that can track the turning sky allows the incoming light to build up on the camera's sensor. The image records nebulae and star fields too faint for the eye to see, revealing a universe below the threshold of human vision.

The Best Cameras

Digital cameras have all but completely replaced film cameras in astrophotography. They offer instant results, an essential aid to nighttime imaging, when exposures must often be guessed at and the subjects can be too faint to frame in the viewfinder. In addition, digital cameras are far more sensitive when shooting dark sky scenes, picking up in seconds or minutes what might have taken hours to record, it succeeded at all.

A glance at the specifications of digital cameras might suggest that sheer megapixel count determines image quality. Not so. For long exposures of faint nighttime subjects, noise, not lack of pixels, is what ruins astrophotos. Electronic noise is present in all digital cameras, and in long exposures can build up to pepper an image with colored, grainy specks.

The consumer cameras best able to capture clean, noiseless images of the night sky are digital single-lens reflex (DSLR) cameras, the kind that allow users to change lenses and look through an optical viewfinder that aims through the same lens that will take the image. DSLRs have larger digital chips with larger individual light-sensitive pixels than do small, pocket-

Steady tripods are essential.

size cameras, despite similar megapixel counts. The larger chips are able to record more light in a given exposure, yielding images with less unwanted noise and more wanted signal.

For best results when shooting long-exposures with DSLRs, there are a few things to consider. Turn on any long exposure noise reduction setting, which can usually be found under Custom Functions or on the Shooting menu. Switching on the High ISO Noise Reduction can also help.

When taking night-sky images the camera has to be held rock steady during the long exposure. That demands a sturdy tripod. Another essential is a remote release, an optional switch that plugs into a jack on the camera and allows the shutter to be triggered without jiggling the camera and to be held open for as long as desired. Extra batteries are also a good idea, as long exposures on cool nights can quickly drain camera batteries.

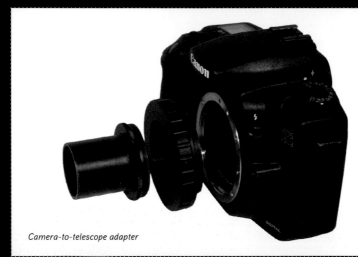

Camera-to-telescope adapter

JUST ABOUT EVERY owner of a camera and a telescope wants to connect the two for unique images. Long exposures of nebulae and galaxies taken through a telescope require complex equipment and techniques best left for a later time when the photographer has gained experience with simpler methods and is willing to pay for the extra gear required. For the first few outings, here are a few techniques to try.

Adapting a Telescope

Any telescope can be turned into a supertelephoto lens for dramatic close-ups of the Moon. The best method is to employ a DSLR camera and remove its lens. In its place goes a camera-to-telescope adapter tube, equipped with a T-ring that allows the adapter to click into the lens mount of the camera just like a lens. The required ring and adapter

are available at any telescope dealer—generic adapters will work with most telescopes, though some telescopes require their own special adapters. A telescope dealer can provide buying advice.

With its adapter in place, the camera then slides into the telescope's focuser in place of the eyepiece and any star diagonal that might be used (in refractors and Schmidt-Cassegrains, for example). Focusing requires care, as the image through the camera viewfinder can be dim. Focus so that the Moon's limb or edge of shadowed craters look as sharp as possible.

Exposures are surprisingly short—which isn't too surprising since the Moon is a bright, sunlit rock. Actual exposures will depend on the telescope and moon phase but are typically 1/15 to 1/500 of a second at ISO 100 to 400. These are short enough that an electric drive on the telescope is not essential, but it is recommended to keep the Moon framed during a shooting session.

While a DSLR camera is best, a point-and-shoot camera (even a cell phone camera!) simply aimed into a telescope eyepiece can grab surprisingly good snapshots of the Moon. Experiment with different techniques to see what results you like best.

Piggyback Shots

A step up in complexity is the realm of "piggyback photography," where the camera tracks the sky during a long exposure. The resulting images can be spectacular, revealing night sky objects in a whole new light.

In this scenario, the camera, equipped with a wide-angle, normal, or short telephoto lens, must ride on the side of a telescope. The telescope itself must have a tracking motor, and it must have an equatorial mount that

is aligned to rotate around the celestial pole. Be sure to check the telescope's manual, which should contain instructions on how to perform the required "polar alignment." The proper alignment is necessary for achieving the pictures you want.

While the setup is more demanding and requires more time and care than just "point and click," the results can be spectacular, especially under a very dark sky. Exposures of two to four minutes at f/2.8 and ISO 800 to 1600 reveal countless stars and faint nebulae in vibrant, rich color, while the stars remain pinpoints because the telescope's tracking system counteracts the streaking effect of Earth's rotation.

After a few successful piggyback shots, you'll be hooked on a lifetime of astronomical photography and the breathtaking images waiting in the night sky.

Camera and telescope together

Star trails twirl overhead.

SOME OF THE most dramatic astronomical photographs are taken with no more than a standard camera on a tripod. The gear and techniques are simple. What makes for a great photo at night is the same that makes any photo a winner—good composition and an eye-catching subject.

Nightfall

After nightfall, moonlight can provide enough illumination to paint the landscape and reveal details. An exposure of 20 to 40 seconds at f/2.8 and ISO 400 produces a scene that looks like daylight, complete with blue sky, but the sky is filled with stars.

Constellations

The beauty of digital cameras, particularly DSLRs, is that they can pick up lots of stars in a relatively short exposure time. Open the lens to about f/2.8 (or with slower "kit" zoom lenses open to wide open at f/3.5 to f/4) and set the camera to ISO 400 to ISO 1600 (for greater sensitivity). Then frame a constellation and open the shutter for 10 to 40 seconds. This is best done from a dark rural site, but even from the city the technique will record bright stars and constellation patterns, though exposures may have to be limited to just a few seconds. A normal or wide-angle lens is best for framing most constellations. From a dark site exposures no more than a minute long at ISO 1600 can pick up the glowing star clouds of the Milky Way.

When the image pops up, you'll be amazed at the vivid colors and detail.

Star Trails

As Earth turns, the entire sky appears to rotate about the celestial pole. In the Northern Hemisphere this point lies near Polaris, the North Star.

Open the shutter for 5 to 30 minutes and the image will record the stars as streaks circling the polestar. The trick is to stop the lens down to f/4 to f/8 and reduce the camera sensitivity to ISO 100. How long the camera can be left depends on the darkness of the sky, the quality of the camera, and the temperature of the night (cold reduces camera noise that might build up during long exposures). But exposures up to one to two hours long are possible with good DSLR cameras.

Taurus's bright stars reveal their colors.

Stars & Constellations

Gemini (upper left) and Orion (upper right) come into view at sunset.

Our Galaxy

GALAXIES—there are tens of billions of them—are among the universe's basic structures, large collections of stars, dust, gas, and mysterious dark matter that are held together by their own gravity and arrayed in clusters around the cosmos. Our galaxy, the Milky Way, is thought to be about 10 billion years old, which is a little younger than its

See "Galaxies," pp. 266-267

SKYFACT

Scientists infer the presence of dark matter by comparing a galaxy's visible matter with its speed of rotation. Typically there is not enough visible material to create the gravitational force to hold the system together, meaning that dark matter—black holes, brown dwarf stars, or an array of exotic objects—must be present. Indeed, dark matter may supply the bulk of a galaxy's total mass.

The Milky Way is a spiral galaxy, like NGC 4414 (above right).

oldest stars, which are around 13 billion years old. Different dating methods have produced different estimated ages, but the Milky Way is apparently among the older galaxies in a universe that began 13.7 billion years ago. The Milky Way's spiral shape is common among larger and brighter galaxies.

The entire spiral structure spins, completing a rotation about every 200 million years. Overall, the Milky

Way has a diameter of about 100,000 light-years. It is thickest in the middle—measuring about 13,000 light-years—while the outer arms of the spiral thin to about 1,000 light-years. Our solar system is located roughly 25,000 light-years from the galaxy core, in what is known as the Orion Arm, about halfway to the outer edge.

Observation

Dark nights and clear skies give the best viewing experience of the Milky Way, which can be easily washed out by even minor sources of light pollution. Earth's ecliptic is tilted nearly 90 degrees to the galaxy, so the view changes seasonally. In the Northern Hemisphere, the summertime perspective is toward the core of the galaxy; the densest region of stars and the brightest in view will be to the south, coursing through Sagittarius. Winter orients the Northern Hemisphere toward the galaxy's outer edge, in Orion and Gemini, a fainter, less star-rich picture. In spring and autumn, we peer beyond the top or bottom of the disk into intergalactic space—a less dense and darker field in which to spot other galaxies.

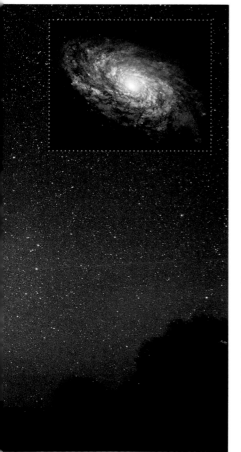

What Is a Star?

CONCEPTUALLY, stars are not that complicated. They are chemically simple, made almost entirely of hydrogen and helium gas. The dynamics behind their formation is not fully understood but is known in basic outline. Though the area between stars and galaxies seems empty, the "interstellar medium" is in fact filled with hydrogen, helium, and dust. It is not spread out evenly but occurs in patches, some of which become dense enough for gravity to begin pulling the material inward. If pressure and heat build adequately at the center of the ball of gas, nuclear fusion is triggered: Protons of hydrogen are fused together into helium, releasing an equivalent of around 0.7 percent of the initial mass as energy. The energy, pushing outward, counteracts the inward pull of gravity and establishes an equilibrium that can last tens of billions of years.

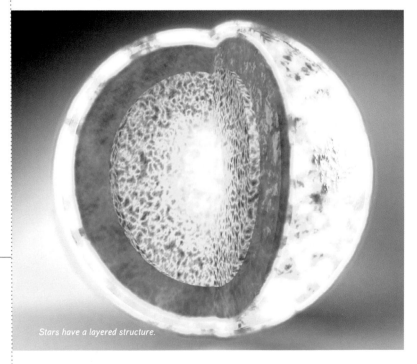

Stars have a layered structure.

SKYFACT

The first star catalogs were put together by Chinese astronomers in the fourth century B.C. Today contemporary lists have cataloged millions of stars brought into view by modern equipment. The Hubble Space Telescope Guide Star Catalog, for example, includes around 19 million stars.

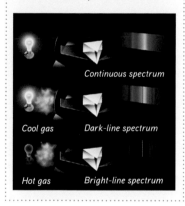

Continuous spectrum

Cool gas   Dark-line spectrum

Hot gas   Bright-line spectrum

Birth

Large gas clouds—literal star nurseries—have been found throughout the Milky Way. One prominent one can be seen by looking through binoculars or a telescope at the middle of the three stars that form the "belt" in the constellation Orion. Several young stars can be seen (with a small telescope) in the Orion Nebula, and studies of the area indicate that others are now

forming. The raw material for star formation can also come from the supernova explosions of older stars across the universe.

Naming the Stars

There are several conventions for naming stars, applied particularly to those included in the major constellations. One of the first—with names that are still used—was created by German astronomer Johann Bayer. In 1603 he published the *Uranometria,* which used the 48 traditional constellations as the basis for his naming system. Combining the Greek alphabet with the constellation names, Bayer used the apparent brightness of the stars to create a list, with the brightest becoming the "alpha" star in each constellation, the second brightest "beta," and on through the 24-letter alphabet (and starting over with capital letters if necessary). Thus, the brightest star in Orion becomes Alpha Orionis (Latin for "the alpha star of Orion"). Bayer's labels aren't perfect—sometimes an alpha star won't be the brightest—but they are commonly found on star guides and charts, and it's a helpful system for defining and discussing the objects under study.

Another system, published about a century later by England's first Astronomer Royal, Rev. John Flamsteed, combined the star's constellation with its relative right ascension, essentially numbering them from west to east. The Latin genitive case is used, so a designation like 1 Orionis would indicate the westernmost star in Orion.

The Stellar Family

TWO SIMPLE traits—color and brightness—were used to create a star classification system.

Henry Draper Catalog

A Harvard University team in the late 1800s began classifying stars based on a spectral analysis of their hydrogen emission lines. Spectral lines, temperature, and color were connected, and then astronomer Annie Jump Cannon, as she combed through tens of thousands of photographs of star spectra, ultimately used a simple color

classification scheme. Today, stars are still classified on the O to M scale she used in compiling the Henry Draper Catalog, a star listing named after the astronomer who began the project. By the end of Cannon's work it included about 225,000 stars.

The scale began as an alphabetical sequence of star spectra. But when Cannon arranged spectra by the strength of their hydrogen lines (which reflected temperature), she found the sequence ran from hot, blue O stars to cool, red M ones, with classifications B, A, F, G, and K in between. More than a century old, the scale has proved an efficient way to convey with a single letter a star's temperature and chemical composition. The cool M-type stars burn at around 5000°F (2760°C),

Hertzsprung-Russell diagram

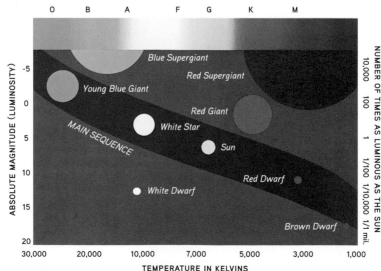

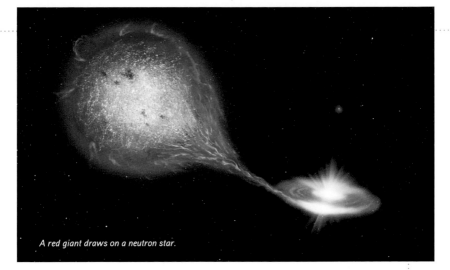

A red giant draws on a neutron star.

for example, and their spectra indicate the presence of metals like magnesium and titanium oxide. O stars burn at more than ten times that temperature and show the presence of ionized helium, carbon, and oxygen. Each category is subdivided further on a 0 to 9 scale based on temperature, from hotter to colder.

Hertzsprung-Russell

Even as Cannon continued her work, Ejnar Hertzsprung of Denmark and Henry Norris Russell in the U.S. began independently plotting the color/temperature of stars against their luminosity, or brightness. The Hertzsprung-Russell diagram, named after both of them, shows an order to the relationship—the hotter, bluer stars are also larger and brighter, while the cooler red dwarf stars are also smaller and dimmer. The H-R diagram shows that most stars lie along what is called the main sequence, a diagonal band that shows the basic connection between the factors temperature, color, size, and brightness.

133

Brightest Stars in the Northern Hemisphere

NAME	CONSTELLATION	SEASON	MAGNITUDE	DISTANCE (LY)
Sirius	Canis Major	Winter	-1.50	8.6
Arcturus	Boötes	Summer	-0.05	37
Vega	Lyra	Summer	0.03	25
Capella	Auriga	Winter	0.08	42
Rigel	Orion	Winter	0.12	773
Procyon	Canis Minor	Winter	0.40	11
Achernar	Eridanus	Winter	0.50	144

Stars Orbiting Stars

A S MORE HAS been learned about stars and their structure, it seems our solo-traveling Sun is an outlier. The massive gas clouds that nurture stars usually produce them in pairs or greater multiples that remain gravitationally attached to each other through their life. Binary stars, triples, and larger groups are the norm, whether the companion is a twin of similar size and luminosity or a group of small siblings that stick close to the dominant member of the system and can be detected only with advanced methods.

Double Stars

Double stars have been recognized for centuries, though it was not until the late 1700s that astronomers began distinguishing so-called optical doubles—coincidental alignments of stars

This triple-star system in Cygnus is about 149 light-years away.

Alpha Centauri double star

began a deliberate hunt for them. Lists of known binary and multiple stars, including the observations of amateur binary hunters like Sherburne Burnham, were eventually compiled into lists like the 17,180-entry Aitken Double Star Catalog.

Seeing Doubles

Powerful binoculars will bring some of the more prominent binaries—Epsilon Lyrae and Beta Cygni for example—into view, but a telescope will be needed to tackle more subtle double stars. Distance, differences in magnitude, and other factors will affect your ability to distinguish the "B" star from its brighter and more dominant "A" companion. Gamma Andromedae will produce an orange-blue pair; a close look at Theta Orionis in the Orion Nebula will yield, depending on the power of the telescope, four to six stars, an interesting configuration known as the Trapezium.

that only appear to be close—from true binary star systems, with two stars orbiting a common center of gravity. In the latter, a telescope is almost always needed to pick out the individual members. Indeed, the partners in one of the most famous pairs of stars in the sky—Mizar and Alcor—are actually about three light-years apart, and apparently are not linked by gravity.

In 1650, however, Italian Jesuit astronomer Giovanni Battista Riccioli—famous for his maps of the Moon and his opposition to heliocentrism—observed that Mizar itself had a close companion (unrelated to Alcor). It was the first discovery of a binary star. Since then, astronomers have found that both components in the Mizar system are also binaries, making Mizar a quadruple star system! (Alcor, however, remains solitary.)

By the late 1770s, enough double stars had been located that English astronomer William Herschel

THE SCIENCE OF Variable Stars

Variable stars change luminosity, sometimes by a large enough degree to determine whether they can be seen with the naked eye. That is the case with Mira in Cetus, which disappears and reappears on an 11-month cycle. "Mira stars" are red giants that have reached an unstable point in their life, and start to pulsate. Rarer Cepheid variables expand and contract on a regular cycle of between one day and five weeks. T Tauri stars are still being formed, and change in luminosity as they contract and shed outer layers of gases. Some of the more prominent to observe include Eta Aquilae and Mira (Omicron Ceti).

Life Cycles of Stars

THE MASSIVE amount of gas contained in a star is, nevertheless, finite: The nuclear furnace at its core will eventually run out of fuel, and the star will die. How long that takes and how it happens depends on the star's size.

Because of the energy needed to sustain them, large stars consume hydrogen at a relatively faster rate than smaller ones do. The Sun, considered a medium-size star, formed with enough fuel to live for about 11 billion years. Far more massive blue giants might burn out in a quick million years or so, while smaller red dwarfs—the most common type of star—will trundle along unchanged for tens of billions of years.

For most of that time, stars live along the main sequence of the Hertzpsrung-Russell diagram, acting in a stable and predictable way. Hydrogen atoms fuse into helium, energy is released, and the outflow of energy offsets the inward tug of gravity—the system is in balance. But as stars near the end of their life, they begin to defy that basic connection and may jump out of the main sequence. A white star, for example, might balloon into a red giant—becoming cooler and redder yet brighter, as it swells enormously in size. Stars like the Sun will eventually collapse into a hot white dwarf.

How a Star Dies

As the fuel tank starts to run dry, the nuclear core begins to shut down. Gravity can no longer be held off, and the star contracts. Temperatures rise, igniting hydrogen in the outer layers of the star and initiating a new round of fusion in which helium atoms (the by-product of hydrogen fusion) are forged into carbon. At this point gravity is overcome by a new surge of energy, and the star begins to expand.

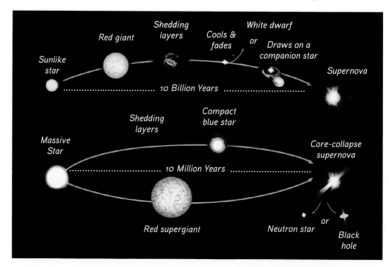

Sunlike star — Red giant — Shedding layers — Cools & fades — White dwarf — or — Draws on a companion star — Supernova

10 Billion Years

Massive Star — Shedding layers — Compact blue star — Core-collapse supernova

10 Million Years

Red supergiant — Neutron star — or — Black hole

Average-size or small objects grow into pulsing red giants (the bright stars Arcturus in Boötes and Aldebaran in Taurus have reached that state). Once the available fuel is consumed, they then shed their outermost layers of gas, exposing the remains of the core and becoming what is called a white dwarf—a spent, stable, and gradually cooling ember. In the case of the Sun, the resulting white dwarf may end up about the size of Earth. The cast-off gas forms a shell-like cloud around the former star.

Going Supernova

More massive stars, with several solar masses of material, have a more violent end. The first few steps are the same, though the larger mass means they expand to supergiant status. In addition, the greater gravitational force present in a contracting core means that temperatures might soar to more than 1 billion degrees Fahrenheit (555.5 million degrees Celsius). Carbon—the product of helium fusion—will itself ignite, starting a process that produces a variety of denser elements, ending with iron. Gravity becomes so intense that subatomic particles fuse in an implosion that compacts the core into an incredibly small space—perhaps 10 miles (16.1 kilometers) across. The resulting shock wave blows apart the outer layers of gas—an explosion called a supernova.

What of the stellar core after these cataclysmic blasts? The dense object left behind either becomes a neutron star—sometimes known as a pulsar for its beaconlike emissions of radiation—or, if large enough, collapses completely and becomes a black hole.

SKYFACT The deaths of old stars in supernova explosions help create new stars, as they throw new elements back into space.

137

NOVAE AND supernovae are the stellar equivalent of atomic bombs—gargantuan explosions. Both are spectacular but caused by different forces.

Novae

Typically a nova occurs when stars in a binary system reach different points in their stellar evolution—with one star collapsed into a dense white dwarf and the other in a red giant phase. If the red giant expands far enough, the white dwarf's gravity may begin siphoning hydrogen from its extended companion. As the cloud of hydrogen accumulates around the dead star and

becomes more massive, pressure and temperature build until the stolen gas erupts in a flash of thermonuclear fusion. The event may brighten the star by as many as ten magnitudes—perhaps bringing it into naked-eye view for a period of months, until the remnants of the blast dissipate.

Novae are not necessarily fatal to the stars involved. In fact, they recur in some binary pairs every few decades,

A supernova remnant in the Crab Nebula

creating a special type of "cataclysmic" variable star that brightens when the explosion occurs, then dims to its original state. Though they are rarely spotted (you can go decades without a sighting), you can hunt for novae with binoculars along the Milky Way.

Seeing a Supernova

Supernovae, by contrast, are onetime events. They can also occur in binary stars if the white dwarf draws in so much hydrogen that its core implodes. Another kind of supernova involves an implosion at the core of a red supergiant star, which unleashes a violent shock wave that blasts away the surrounding cloud of gas. Supernovae unleash so much energy that the star involved briefly shines as bright as its entire home galaxy. Supernovae can occasionally be seen by the naked eye—events that have astonished earthbound observers when a bright new star suddenly appears in the sky. First recorded in Chinese records from A.D. 185, supernovae were also noted in 1006 (the brightest on record, at magnitude -7.5), in 1054, in 1572, and most recently in 1987 in the nearby galaxy known as the Large Magellanic Cloud. The Crab Nebula (M1) in Taurus is what remains of the 1054 event, a thousand years after the fact.

Supernovae can be spotted in your backyard with amateur equipment. But as a practical matter it will require an 8-inch or larger telescope and the skill to train it on other galaxies in search of "new" stars not included on any chart or guide.

An energy-releasing black hole

Black Holes

Although some supergiant stars collapse into pulsars at the end of their lives, the largest supergiants produce such massive gravitational forces that matter literally folds in on itself and collapses to form a black hole. Among the universe's truly strange and awesome events, black holes are objects in which the distance between subatomic particles has been reduced to zero, and where both gravity and density expand toward infinity. Nothing can escape, not even light—hence the name "black hole."

SKY WATCHERS: Tycho Brahe

Born in 1546, Tycho Brahe charted the positions of hundreds of stars. In late 1572 he noticed what he dubbed a new star in Cassiopeia. When Brahe watched as the "new star" faded over 18 months from its peak -4 magnitude, he realized that the object undercut the notion that stars were unchangeable. We now know that he was observing a supernova—proof that even the stars are temporal.

Clusters & Nebulae

D EEP SKY TERRITORY is the domain of star clusters and nebulae, dramatic formations sometimes visible to the naked eye but which under the amplification of a telescope give a deeper sense of the universe's scale and complexity. This perspective of the sky is breathtaking.

Star Clusters

Star clusters are of two basic sorts. Open, or galactic, clusters, are looser concentrations of anywhere from a few to several thousand stars that shared a common birth cloud. Groupings like the Pleiades, in the constellation Taurus, have been recognized for centuries. Globular clusters also share a common origin but are several orders of magnitude larger. These massive collections of stars—as many as a million—are typically around 13 billion years old. About 150 globular

Globular cluster Omega Centauri

Keyhole Nebula

clusters have been identified, mostly in the outer reaches of the Milky Way. They can be spotted best around the constellations Sagittarius and Ophiuchus. Naked-eye observation is possible, but a 4-inch telescope will begin to isolate stars, while a 10-inch instrument will show a grainy profusion.

Nebulae

A nebula is a cloud of stellar gas, and while the dramatic images in astronomy books is out of reach of most backyard equipment, they still give a sense of the universe's dynamics. Emission nebulae are star-forming clouds set aglow by the energy of the young stars within them—look to the middle star of Orion's belt. Planetary nebulae are actually the gas being blown away from a dying red giant, such as the Ring Nebula in Lyra. Dark nebulae are collections of interstellar dust that hide the light of the stars behind them— one area near the Southern Cross has been dubbed the Coalsack. Reflection nebulae shine from the light of nearby stars. Many nebulae are visible in a 4- to 6-inch telescope. Use higher magnification for better contrast, and experiment with light pollution filters to enhance visibility.

Clusters

NAME	MESSIER NUMBER	CONSTELLATION	VIEW	TYPE
Pleiades	M45	Taurus		Open cluster
Beehive	M44	Cancer		Open cluster
Coma Berenices	none	Coma Berenices		Open cluster
Great Globular Cluster	M13	Hercules		Globular cluster
Wild Duck Cluster	M11	Scutum		Open cluster

Constellations & Asterisms

I N THE CAVES of Lascaux, France, prehistoric frescoes drawn some 16,000 years ago include what researchers believe are elements of an early sky map. There are what appear to be diagrams of the phases of the Moon, references to the Pleiades star cluster, and a representation of the three-star group known as the Summer Triangle. It is evidence of a human urge to order the heavens—whether for purposes of tracking the seasons, guiding travel and navigation, or conveying history through stories. The process has been handed down through the patterns known as constellations and asterisms. The difference between the two is important.

Constellation Origins

There were 48 constellations identified in antiquity, including the large

THE STORY OF Orion

Among the most prominent and literally drawn of the constellations, Orion—in Greek lore a hunter slain mistakenly by Artemis, goddess of the hunt—stands out in the northern winter, anchored by the bright stars Rigel and Betelgeuse, and including the unmistakable, three-star belt across the middle. The image is so intuitively that of a large person with a raised hand that many different cultures have given the group of stars a similar interpretation. They were called Al Jabbar (the Giant) by Syrian astronomers, and Sahu (the soul of the god Osiris) by the Egyptians. Thousands of miles and an ocean away, Native Americans in New Mexico referred to the star group as Long Sash, a heroic figure from their own mythology.

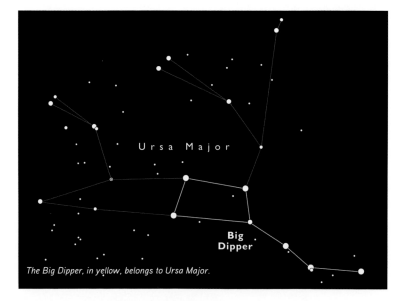

Ursa Major

Big Dipper

The Big Dipper, in yellow, belongs to Ursa Major.

and more well-known star patterns like Orion, Scorpius, and Pegasus. Incorporated into poems like Aratus's "Phaenomena" in the third century B.C., and codified in Ptolemy's *Almagest* in the second century A.D., the names in use today reflect their origins in Greco-Roman mythology.

Yet by giving a larger identity to individual stars, particularly the brighter ones in the sky, constellations have helped create a map of the heavens that is still used today. Beginning in the 1500s, more frequent travel to the Southern Hemisphere brought into view areas of the sky unknown to the Babylonians, Greeks, or Romans, and with that came a whole new set of constellations. Ultimately the list grew to include 88 star patterns, with a handful of new northern ones added, some proposed by different celestial cartographers but dropped, and the whole of the south celestial polar region filled in.

During the 1920s, with the pace of discovery in astronomy quickening, the International Astronomical Union (IAU) formalized the boundaries of each constellation so that each represented not just a group of stars but a roughly rectangular patch of sky. That constellation-driven map is now a basic tool for astronomers to orient themselves when looking for variable

The Big Dipper

stars, clusters, galaxies, or other deep sky objects: All will reside in one of the bordered areas sanctioned by the IAU.

Asterisms

Don't be surprised when you look at the list of constellations and find some prominent names, like the Big and Little Dippers, missing. These are asterisms, small groups of stars that have a distinct, well-known shape and may form part of a constellation but are not considered constellations in and of themselves. The dippers, for example, are part of Ursa Major and Ursa Minor. Useful for finding your way around the sky, other prominent asterisms include the Summer Triangle formed by the bright starts Altair, Deneb, and Vega; the W (or M) shape in Cassiopeia; and the sickle that forms the head of the lion in Leo.

143

SKYFACT

The traditional list of constellations began to expand in 1595, when Dutch navigator Pieter Keyser, on an expedition to the East Indies, suggested a dozen new star formations, all added to Johann Bayer's 1603 atlas of the heavens, the *Uranometria*. The trip would have taken him around the tip of Africa, at roughly 34.5° south latitude.

Northern & Southern Skies

WHEN PTOLEMY codified the constellations, he was using traditions most likely handed down from Mesopotamian observers who had watched the skies at about 35° north latitude. The northern sky up to the celestial north pole would have been visible. Theoretically they could have looked over a 90-degree span to the south as well—or down to a declination of around -55. The practical limit would be perhaps ten degrees less than that since objects near the horizon are difficult to see. Anything farther south would have been blocked by Earth.

That partly accounts for why the ancient astronomers limited their constellation list to 48. Some southern features like Crux, the famous Southern Cross, crept over the horizon, but not enough to make a full impression: Until the 16th century it was included as part of Centaurus.

> **SKYFACT**
> The Southern Cross is the smallest constellation, yet is so dominant in the southern sky that it appears on the flags of Australia and New Zealand.

Your Viewpoint

Likewise, your view of the constellations will be determined by your latitude. Though constellations are often characterized as belonging to the

See "Getting Oriented," pp. 14-15

Backyard Guide to the Night Sky

The southern celestial sky

The Coalsack nebula is a southern sky highlight.

northern or southern hemisphere, it helps to envision them on a continuum. Using the celestial coordinate system, each constellation occupies a range of declination—akin to terrestrial latitude—on a hypothetical celestial sphere that encases Earth. Those with more northern declinations on the 0-to-90-degree scale will be higher and more visible for more of the year in the northern hemisphere (likewise in the south for those with negative, southerly declinations from 0 to -90).

Indeed, at the extreme, some constellations, like Ursa Minor, are "north circumpolar"—meaning they are at such a high northern declination that they are visible in the northern hemisphere throughout the year, never dipping below the horizon. Some of these may peek into the south as well during certain

THE STORY OF Argo Navis

French astronomer Nicolas-Louis de Lacaille, after a four-year expedition to the Cape of Good Hope in the 1750s, made a major revision to the largest constellation at the time, Argo Navis, which represented a mythical ship. Lacaille broke it into component parts—Carina (the Keel), Vela (the Sails), Puppis (the Stern), and Pyxis (the Compass).

times of the year, but just barely, visible only to people within a few degrees of the Equator.

The bulk of the constellations are not circumpolar. Some prominent constellations, Orion and Virgo, for example, are positioned so close to the celestial equator that they can be seen from both hemispheres on a seasonal basis In between the circumpolar constellations and the equatorial ones, what you see depends on where you are.

The Zodiac

The astrological clock in Prague

A S CONSTELLATIONS were added to Ptolemy's list, it turned into a grab bag of items, a compendium not just of myths and monsters but of microscopes and sextants and other objects that reflected the scientific sensibility of the 17th and 18th centuries. But the most ancient star patterns remain at the core and form a sort of foundational first step to learning the sky. These are the constellations of the zodiac: the 12 well-known animals, people, and creatures that appear in an annual cycle along a path tightly bound to the ecliptic. The names of the zodiacal constellations are known to almost anyone, tied up as they are in the pseudoscience of astrology and bound to pop culture since at least the dawning of the age of Aquarius.

Star Signs

The word "zodiac" comes from the Greek *zodiakos kyklos,* or "circle of animals." Most do have animal associations: Aries (the Ram); Taurus (the Bull); Gemini (the Twins); Cancer (the

SKYFACT

When the Tropics of Cancer and Capricorn were associated with the summer and winter solstices, it was because the Sun was "in" those constellations at the time. Now, because of precession, the sun is in Sagittarius during the winter solstice and in between Gemini and Taurus during the summer.

Crab); Leo (the Lion); Virgo (the Virgin); Libra (the Scales); Scorpius (the Scorpion); Sagittarius (the Archer, a centaur); Capricornus (the Sea Goat); Aquarius (the Water Bearer); and Pisces (the Fish).

Some of the original associations have been lost to history. The twins in Gemini are usually linked today with Castor and Pollux of Greek mythology but have also been pairs of peacocks and goats. Capricornus may have started as a simple goat—a common sight to Chaldean or Babylonian herdsmen—but "grew" a fish tail, a possible reference to a Greek tale in which the god Pan flees a monster by jumping into the Nile.

Phases

What truly sets the zodiac apart from other constellations is its position along the ecliptic, the path of Earth's orbit around the Sun. Hence the illusion of a wheel: Over a year, as the planet moves through its orbit, each of the 12 will follow in succession, coming into view in the east, rising higher night by night, then lowering into the west until that patch of sky has disappeared for the year.

It is this westernmost phase, when the Sun is roughly aligned with—or "in"—a particular constellation, that is associated with the "signs" of the zodiac. Sagittarius, for example, after spending the northern summer months prominent in the night sky, has by late fall shifted to the west. The Sun is roughly in front of it, and by twilight in the north it has almost

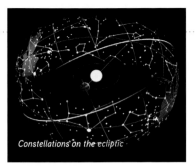
Constellations on the ecliptic

disappeared below the horizon. It is felt to "control" the period from November 22 to December 21.

THE STORY OF Astrology

Virtually everyone knows their "sign"—the constellation of the zodiac where the sun resided when they were born. The practice of astrology involves predicting the future based on the alignment of the Sun, Moon, stars, and planets at a particular point. Around as long as stargazing, it is an unfounded bit of mock science spun from otherwise reasonable observations. The Egyptians looked to the rising of Sirius, for example, to predict the flooding of the Nile, and the events did coincide. However, one did not cause the other, a leap that astrologers readily insist on making. They further complicate the calendar by dividing it into spheres of influence for each constellation. At the largest scale, the 25,800-year precession of the equinoxes—the time it takes to place the spring equinox in each of the 12 constellations of the zodiac—gives each constellation a ruling "age" (depending on the calculations and the borders set for the constellations, we are either in the age of Aquarius or nearing the end of Pisces).

SkiesThrough the Year

LEARNING THE constellations will take patience and a willingness to work around different obstacles. But it requires no equipment—far from it. Remember, these are star patterns first identified by the naked eye thousands of years ago. They can occupy a large portion of the sky, so a wider field of view is essential. But the circumstances of a Mesopotamian farmer are not what we face today. A viewing place too close to the city, for example, will wash out important ref-erence stars with light pollution. A visit to dark-sky country, by contrast, may leave a beginner bewildered trying to pick out unfamiliar reference points from a crowded backdrop.

Constellation Calendar

One method is to build slowly with what is easily available, while paying attention to the calendar. From a typical urban location, for example, you may find the sky in one or more directions blocked by backyard trees or nearby buildings. So determine which vista is open, and locate sky charts that will show you what's in that part of the sky over the course of the year. In the mid-northern latitudes, with a reason-

Backyard viewing changes with the seasons.

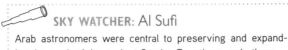

SKY WATCHER: Al Sufi

Arab astronomers were central to preserving and expanding the work of the ancient Greeks, Egyptians, and others,
helped by patronage from the great rulers in Baghdad and
Persia. Among them was Abd al-Rahman al-Sufi, a tenth-
century Persian astronomer who also worked in Baghdad. Al-Sufi supplemented Ptolemy's *Almagest* with
his own estimates of stellar magnitude and
other observations, creating his *Book
on the Constellations of Fixed Stars.*
Included was what he called the
"little cloud" around Andromeda—
now regarded as the first recorded
observation of the Andromeda galaxy.

able portion of the southern sky open, for example, you can still enjoy the summer views of Sagittarius and Scorpius, then watch as Orion begins to dominate the sky as fall progresses.

Be sensitive to the cyclical aspect of the constellations. While the few that are circumpolar will theoretically always be visible, they will also, depending on latitude, dip lower in the sky over the course of the year. They, too, have a season when they are highest and most visible.

To plan your year, a sky chart is essential. Computer programs will allow you to pick a date and time and set coordinates, as will websites like Astronomy.com. Monthly magazines, will also do the trick, as will a variety of books and other publications.

Starry Signposts

The brighter stars in the larger constellations will provide the best beginning reference points. Typically, charts

will use Bayer designations—a Greek letter and a Latin constellation name— to designate a constellation's component stars in order of magnitude. But not all "alpha" stars are created equal. The reason Orion stands out is that it includes three of the brightest stars in the sky—Rigel, Betelgeuse, and Bellatrix. Canis Major includes Sirius, the very brightest, as a reference point. Gemini has the bright stars Castor and Pollux for reference points. Most likely, these will be indicated on your chart by larger circles. Look for the biggest, determine when those constellations will come into your field of view, and spend some time with them.

Then you'll be prepared to star hop to fainter star patterns. Even some of the signs of the zodiac require proper orientation. Cancer, for example, has no star brighter than 4th magnitude. But it rests between Gemini and Leo, two show-offs with bright stars that can help locate the scuttling crab.

THE NIGHT SKY is full of wonders, and those wonders change over the course of the year, as the constellations and other star patterns slowly make their way across the heavens. To help you get started in your explorations, let us give you a quick tour of five well-known and easy-to-find star patterns: One for each season of the year, and one more you can see all year-round.

The Big Dipper

The Big Dipper is probably the most familiar star pattern in the entire northern sky, in part because it's so far north that most people in the Northern Hemisphere see it all year long. But what many people don't know is that it's not actually a constellation.

Rather, it's what known as an asterism, a pattern of stars that is not one of the 88 constellations officially recognized by the International Astronomical Union. Asterisms can be comprised of stars entirely within an already recognized constellation, or of stars from more than one constellation. In the case of the Big Dipper, it's contained within an official constellation: Ursa Major, which is Latin for "Great Bear." The handle of the Big Dipper is located in the Great Bear's tail, while the square body of the dipper is the bear's hindquarters.

Other Cultures

While we here in the United States call this asterism The Big Dipper, it has been called many things by many cultures over the years—a reminder that not every culture has seen the same thing in the night sky, even while looking at the same pattern of stars. In England and Ireland they called it The Plow or The Starry Plow; while in Scandanavia and Germany it's seen as a wagon. France sees it as a saucepan. Hindu astronomers called it Saptarshi, or the Seven Great Sages. It even makes the translation in to fictional worlds, most prominently in J. R. R. Tolkien's the *Lord of the Rings*, where it is seen by the characters as the Sickle of the Valar. The Big Dipper is familiar enough that it even shows up on the state flag of Alaska, as a symbol of that state's northerness.

Big Dipper and Little Dipper

Reference Point

The Big Dipper's status as an asterism is appropriate because in many ways it can be used as a guide to find other stars and celestial objects. For example, extending north the imaginary line between Merak and Dubhe leads you to Polaris, better known as the North Star. Following the curve of the Big Dipper's handle outward leads to Arcturus in the constellation Boötes (and the third brightest star in the northern sky); continuing on the curve leads to Spica, the brightest star in the constellation Virgo. There's even a saying for this among stargazers: "follow the arc to Arcturus and speed on to Spica."

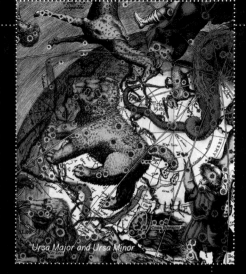

Ursa Major and Ursa Minor

Objects in the Dipper

There are also plenty of fascinating things to see in and around the Big Dipper itself. Start with Mizar, the second star in the dipper's handle. If you look closely at Mizar, you might see that there's another, less bright star very close to it. That star is Alcor, and the two stars together form what is known as an optical binary, which is when two stars that look close together to us but which don't in fact form a true binary system (Alcor and Mizar are light-years away from each other). These two stars also form their own asterism known as the Horse and Rider. For a number of cultures, including the ancient Romans, the ability to distinguish the two stars separately was used as a test of visual acuity. Also near the Big Dipper is the Pinwheel galaxy (more formally known as M101), which is visible with both the help of a small telescope and good seeing conditions.

Orion

THE WINTER and spring skies are populated by some of the brightest stars in the night sky. Also appearing in the heavens are two of the most recognizable and popular constellations, Orion and Leo.

Orion in Winter

In the Northern Hemisphere, Orion is the lord of the winter sky, his distinctive shape filled with bright stars and other astronomical sights. The constellation is named for a famed hunter of Greek mythology, but it is not the only story associated with this star pattern. Other cultures have seen the constellation as representing a shepherd or a harvesting scythe, because it first appears in the northern sky during harvest times. It's also one of the few constellations referenced in the Bible, most notably twice in the Book of Job (9:9 and 38:31).

Orion features two of the brightest stars in the sky. To the north, at the constellation's shoulder, is Betelgeuse, the ninth brightest star; to the south is Rigel, the sixth brightest. Both stars are fascinating. Betelgeuse's diameter is larger than the orbit of Earth, and has 20 times the mass of the Sun. It is a young star, but its mass suggests that it won't live much longer (astronomically speaking) and might undergo a supernova explosion. Rigel is also quite large (17 solar masses) and, thanks to its proximity to the Equator, was one of the "nautical stars" that sailors would use to locate themselves on the ocean. But the real action, astronomically speaking, is in Orion's belt and the "sword" that hangs from it. There you will find the Orion Nebula, one of the few nebulae easily seen with the naked eye (although its real beauty requires a telescope).

Leo

Leo in Spring

The star pattern Leo, one of the 12 constellations of the zodiac, has been consistently seen as a lion across several cultures, including the ancient Egyptians, Persians, Babylonians and Greeks. The most famous story associated with the constellation comes to us from the Greeks, for whom the constellation represented the Nemean Lion, whom the hero Hercules was ordered to kill as the first of his 12 labors.

A notable feature of Leo is an asterism known as the Sickle, a curved set of stars that represents the lion's head and mane, and terminates with Regulus (The Little King), the brightest star in the constellation. Regulus is also the brightest star closest to the actual plane of the ecliptic and as a result is frequently occulted by the Moon. It is less frequently occulted by planets, but if you happen to be looking at Regulus on the night of October 1, 2044, Venus will cross directly in front of the star. Regulus is also unusual in that it has an extremely fast rotation for its size: It is three and a half times the mass of our Sun but rotates every 15.5 hours (by comparison, it takes a month for the Sun's equator to do a full rotation). The star is actually noticably flattened, and its poles are hotter than its equator, enough so that the star is actually darker at the equator (this is known as gravity darkening) If the star spun much faster, it would literally tear itself apart.

SUMMER AND autumn skies have two easy-to-spot sets of stars, the Summer Triangle and Cassiopeia.

The Summer Triangle

Just because all of the officially recognized constellations were in place by the 18th century doesn't mean that people aren't still finding new pictures and patterns in the sky. The Summer Triangle—an asterism featuring three bright stars in three separate constellations—is an example of this: While the asterism itself was first noted in the 19th century, the name Summer Triangle was not popularized until the 1950s, when British broadcaster and astronomer Sir Patrick Moore used it, and astronomer H. A. Rey used it in his guidebook *Find the Constellations.* The stars that create it are three of the brightest in the northern sky. They are Vega, located in the small constellation Lyra (the Lyre); Deneb, in Cygnus (the Swan); and Altair, in Aquila (the Eagle). In the summer months, these three stars are visible right overhead for most people in the Northern Hemisphere.

Each of these stars also has its own story to tell. Vega is the second brightest star in the northern sky (only Arcturus is brighter, and Sirius is in the southern half of the sky). Vega is close to us (some 25.3 light-years away) and is 37 times more luminous than our own star. On July 17, 1850, Vega was the first star, other than the S███ be photographed. Deneb is far███ away (approximately 1,550 light-years) but

. Deneb

Vega

Altair

Summer Triangle

it is estimated to be 60,000 times more luminous than the Sun. Altair is "only" ten times more luminous than the Sun and is closer to Earth—just 16.7 light-years away, and lik █ Regulus, it's a star whose rate rotation is so quick that the entire star is flattened at the poles. Because the star is close enough to be photographed, it was the first on which gravity darkening was directly observed.

Cassiopeia

Cassiopeia

Cassiopeia is a constellation with a distinctive shape that makes it easy to spot in the fall northern sky: It's shaped like the letter M (or W), with one limb of the letter slightly askew, with each endpoint of a line of the letter noted by one of the five stars of the constellation. Of course, not every culture saw this configuration as an M or as a queen. Early Arabian cultures saw a hand, with each star representing a finger.

One notable thing about Cassiopeia is that as a constellation, it lies in the path of the Milky Way galaxy, to which our Sun and all the individual stars we see in the night sky belong. What this means is that when you are looking toward Cassiopeia, you are looking in to our galaxy, and toward the vast majority of billions and billions of other stars that reside in it with us. Another notable thing that distinguishes Cassiopeia, although it's

something you can't see from Earth, is how it would look from space. If you were to travel to Alpha Centauri, the closest star system to our own, and you wanted to find the Sun, you would look toward the constellation Cassiopeia, which would still be recognizable even in space. Our Sun would be there in it, and shine as one of the brightest stars in that night sky.

7

Four Seasons of the Sky

In this chapter:
The Sky Charts
Winter
Spring
Summer
Autumn

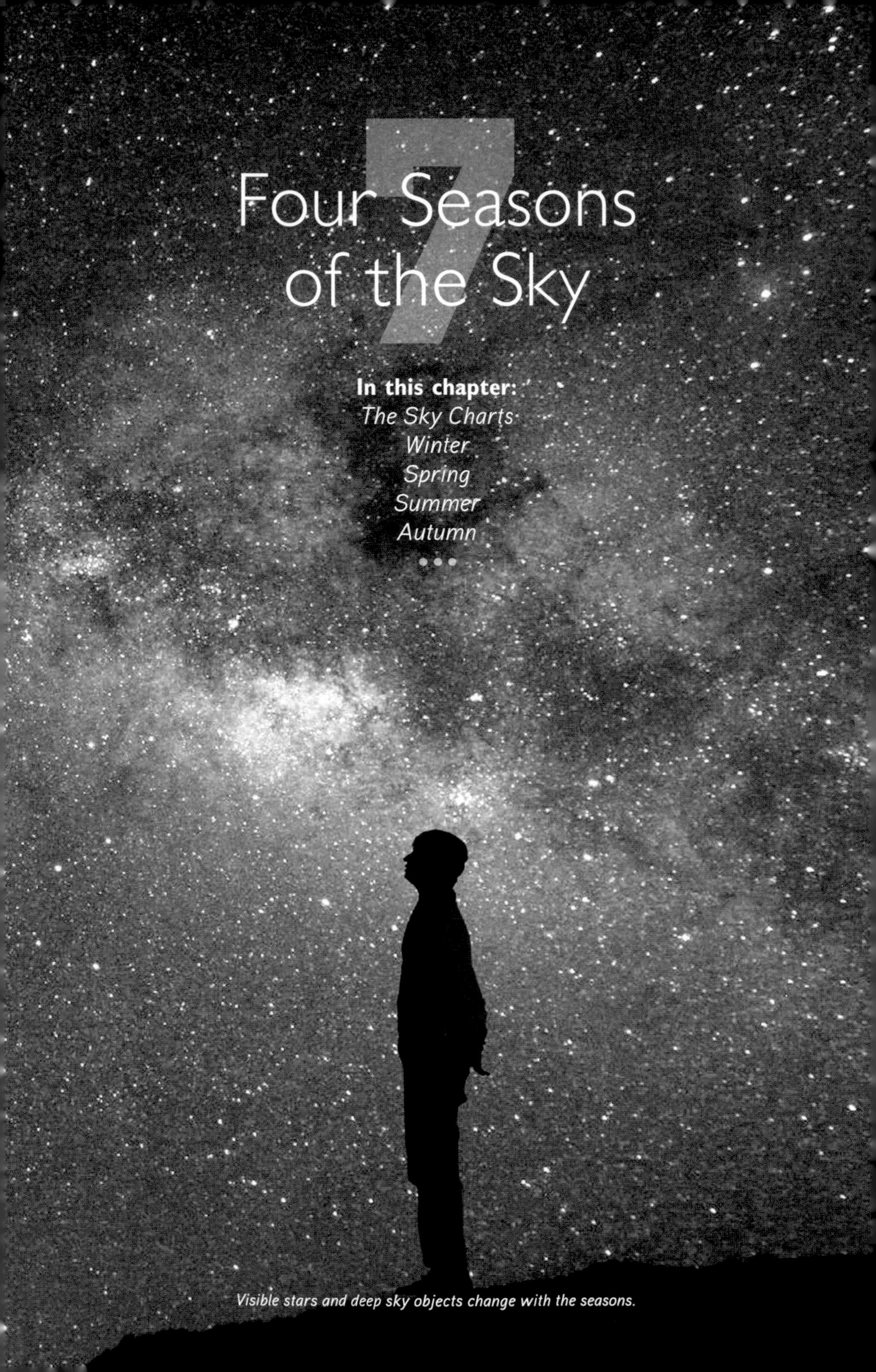

• • •

Visible stars and deep sky objects change with the seasons.

The Sky Charts

THE FOLLOWING eight pages constitute a basic guide to the sky in the Northern Hemisphere. Your view is divided over a series of four maps.

The maps are organized by season—there is one each for winter, spring, summer, and autumn—and they show the visible constellations, the brightest stars, and the locations of the more accessible galaxies, nebulae, and other deep sky objects.

Remember, the celestial sphere operates on a constantly changing continuum: The positions shown apply to particular days, times, and latitude. Specifically, the maps are prepared from the perspective of someone at 40° north latitude, and the dates and times are indicated on each chart.

STELLAR MAGNITUDES

● -0.5 and brighter	● 2.1 to 2.5
● -0.4 to 0.0	● 2.6 to 3.0
● 0.1 to 0.5	• 3.1 to 3.5
● 0.6 to 1.0	• 3.6 to 4.0
● 1.1 to 1.5	· 4.1 to 4.5
● 1.6 to 2.0	· 4.6 to 5.0
	⊛ Variable star

DEEP SKY OBJECTS

⊛	Open star cluster
⊕	Globular star cluster
☐	Bright nebula
✧	Planetary nebula
⬭	Galaxy

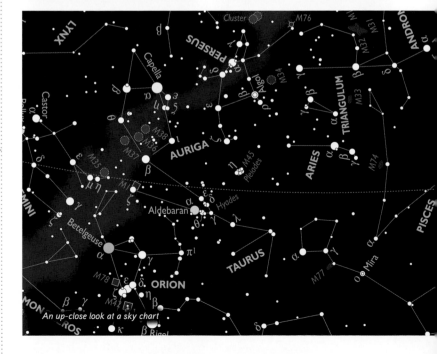

An up-close look at a sky chart

Best Viewing Times

CONSTELLATION	STAR CHART	MONTHS
Auriga	Winter	Dec. / Jan.
Boötes	Summer	June
Cancer	Spring	Mar. / Apr.
Capricornus	Summer	Aug. / Sept.
Cygnus	Summer	Aug. / Sept.
Lyra	Summer	July / Aug.
Orion	Winter	Jan. / Feb.
Pegasus	Autumn	Sept. / Oct.
Sagittarius	Summer	July / Aug.
Taurus	Winter	Jan. / Feb.

Getting Oriented

These maps are an approximation of what you'll see at a particular time but are still useful any night. Like the Sun, the stars and constellations appear to move from east to west, in some cases falling from view below the horizon for several months.

Observers at a latitude above 40° N will find the northern constellations higher in the sky, for longer periods of time, while losing sight of some constellations below the southern horizon. The reverse is true as a position moves toward the Equator. To illustrate, take a look on the maps at the North Star, Polaris, which does not rise and set or change position during the year. It is roughly in the middle of the northern sky—reflecting the orientation of 40° north latitude. Observers farther north will see it higher in the sky. By contrast, it will be lower in the northern sky as your point of observa-

tion moves farther south. Other stars will shift accordingly.

Features

The maps include a silhouette line around the border indicating the horizon. Although constellations near the horizon are visible hypothetically, anything within about ten degrees of that edge will be difficult to spot.

Stars to magnitude 5 have been connected by lines to form the constellations. Stars of magnitude 3.5 or brighter have been labeled for reference, and some have been tinted to reflect their color. The stars' size reflects their magnitude. Deep sky objects are indicated with different symbols (displayed in the accompanying legend). The ecliptic—Earth's path around the Sun, and the line of travel for the zodiacal constellations—is the dotted line across each map; the faint white field represents the Milky Way.

Winter

Betelgeuse sits on Orion's left shoulder, and Rigel forms the right foot. At the sword you'll find the Orion Nebula. Just to the northwest of Betelgeuse is Taurus and the two famed star clusters, the Pleiades and the Hyades.

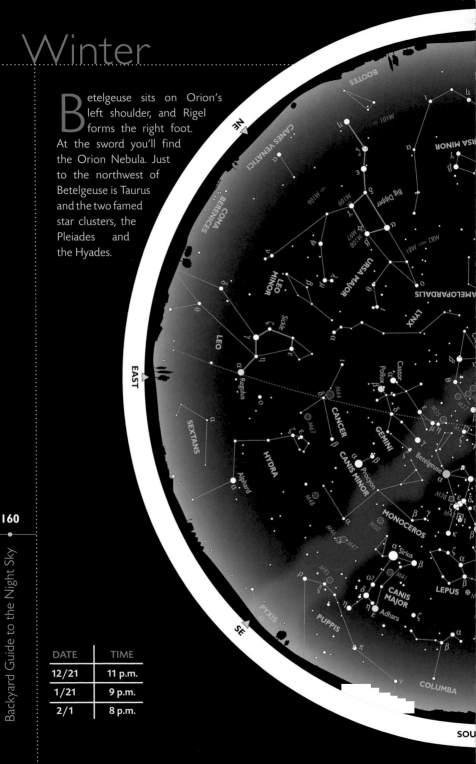

Backyard Guide to the Night Sky

DATE	TIME
12/21	11 p.m.
1/21	9 p.m.
2/1	8 p.m.

Auriga is located by drawing a line almost directly north to the bright star Capella. To the northeast of Orion, Gemini is marked by the stars Castor and Pollux. Sirius is to the southeast and belongs to Canis Major. North of Sirius at the bright star Procyon is Canis Minor.

STELLAR MAGNITUDES

- −0.5 and brighter
- −0.4 to 0.0
- 0.1 to 0.5
- 0.6 to 1.0
- 1.1 to 1.5
- 1.6 to 2.0
- 2.1 to 2.5
- 2.6 to 3.0
- 3.1 to 3.5
- 3.6 to 4.0
- 4.1 to 4.5
- 4.6 to 5.0
- Variable star

DEEP SKY OBJECTS

- Open star cluster
- Globular star cluster
- Bright nebula
- Planetary nebula
- Galaxy

Spring

Spring is galaxy-hunting season. Earth's position relative to the Milky Way has shifted so that the core of the galaxy lies near the horizon in the east. We look out into deep space with a better view of elusive night sky objects.

DATE	TIME
3/21	11 p.m.
4/1	10 p.m.
4/21	9 p.m.

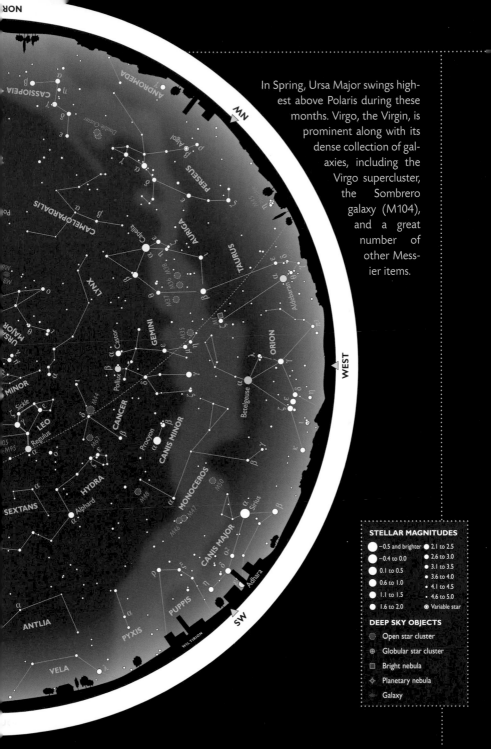

In Spring, Ursa Major swings highest above Polaris during these months. Virgo, the Virgin, is prominent along with its dense collection of galaxies, including the Virgo supercluster, the Sombrero galaxy (M104), and a great number of other Messier items.

STELLAR MAGNITUDES

- −0.5 and brighter
- −0.4 to 0.0
- 0.1 to 0.5
- 0.6 to 1.0
- 1.1 to 1.5
- 1.6 to 2.0
- 2.1 to 2.5
- 2.6 to 3.0
- 3.1 to 3.5
- 3.6 to 4.0
- 4.1 to 4.5
- 4.6 to 5.0
- Variable star

DEEP SKY OBJECTS

- Open star cluster
- Globular star cluster
- Bright nebula
- Planetary nebula
- Galaxy

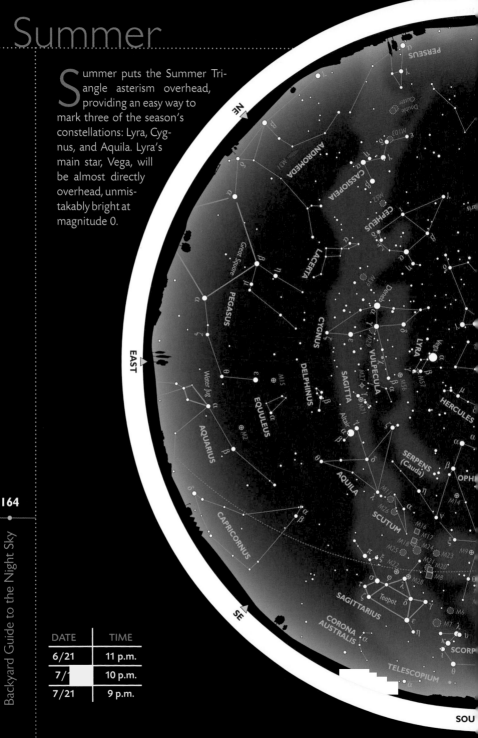

Summer

S ummer puts the Summer Triangle asterism overhead, providing an easy way to mark three of the season's constellations: Lyra, Cygnus, and Aquila. Lyra's main star, Vega, will be almost directly overhead, unmistakably bright at magnitude 0.

DATE	TIME
6/21	11 p.m.
7/	10 p.m.
7/21	9 p.m.

Look near Sagittarius, an area rich in star clusters and nebulae, for some of summer's highlights. Locate the Teapot asterism at its center to spot the notable objects surrounding its lid: the Lagoon (M8) and Trifid (M20) Nebulae, and the Great Sagittarius star cluster (M22).

STELLAR MAGNITUDES

● −0.5 and brighter	● 2.1 to 2.5
● −0.4 to 0.0	● 2.6 to 3.0
● 0.1 to 0.5	● 3.1 to 3.5
● 0.6 to 1.0	• 3.6 to 4.0
● 1.1 to 1.5	• 4.1 to 4.5
● 1.6 to 2.0	• 4.6 to 5.0
	⊙ Variable star

DEEP SKY OBJECTS

⊙ Open star cluster

⊕ Globular star cluster

☐ Bright nebula

✦ Planetary nebula

— Galaxy

Autumn

The Great Square of Pegasus, an asterism at the center of the constellation of the same name, is central to locating several other autumn constellations, including Andromeda. Alpheratz is shared by Andromeda and the asterism.

DATE	TIME
9/21	11 p.m.
10/21	10 p.m.
11/1	8 p.m.

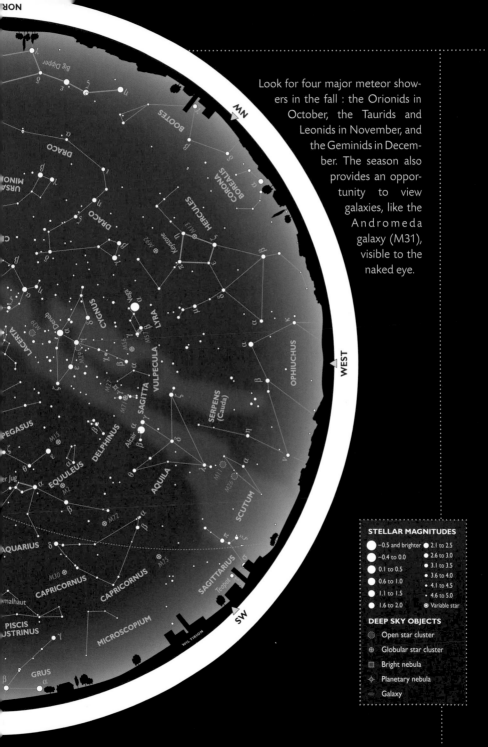

Look for four major meteor showers in the fall : the Orionids in October, the Taurids and Leonids in November, and the Geminids in December. The season also provides an opportunity to view galaxies, like the Andromeda galaxy (M31), visible to the naked eye.

STELLAR MAGNITUDES

- −0.5 and brighter
- −0.4 to 0.0
- 0.1 to 0.5
- 0.6 to 1.0
- 1.1 to 1.5
- 1.6 to 2.0
- 2.1 to 2.5
- 2.6 to 3.0
- 3.1 to 3.5
- 3.6 to 4.0
- 4.1 to 4.5
- 4.6 to 5.0
- Variable star

DEEP SKY OBJECTS

- Open star cluster
- Globular star cluster
- Bright nebula
- Planetary nebula
- Galaxy

Constellations

An 18th-century depiction of the Copurnican System of the universe

The Constellation Charts

THE PAGES that follow are your road map to the stars—a more detailed look at 58 of the 88 recognized constellations. The original 48 described in Ptolemy's *Almagest* are included, as are newer constellations that were added later to "fill in" the sky and that are visible in the Northern Hemisphere. Those not included are the "deep south" constellations that cannot be viewed from the mid-northern latitudes.

Like any guide to the stars, this one has an orientation. The months for best viewing are the times of year when a given constellation will appear highest in the sky during the evening hours for observers at 40 degrees north latitude. Each constellation's entry will refer back to the four seasonal charts in the previous chapter.

A variety of other information is included for each constellation.

❶ FAST FACTS

Underneath the name of each constellation, you'll find a short list of key items and facts. "Makeup" refers to the number of brighter stars (down to magnitude 5) found in the particular constellation—all visible to the naked eye under dark skies and good viewing conditions. "Best Viewed" indicates the best months of the year to hunt for this constellation. "Location" will direct you to the appropriate seasonal

Backyard Guide to the Night Sky

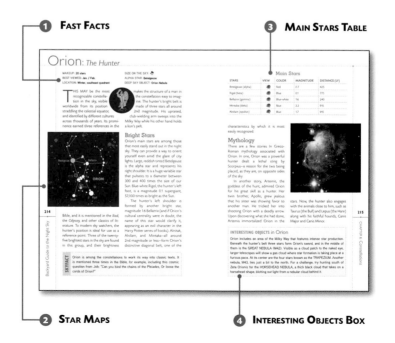

❶ **FAST FACTS**

❸ **MAIN STARS TABLE**

Orion: *The Hunter*

MAKEUP: **20 stars**
BEST VIEWED: **Jan. / Feb.**
LOCATION: **Winter, southeast quadrant**

SIZE ON THE SKY: ⚹
ALPHA STAR: **Betelgeuse**
DEEP SKY OBJECT: **Orion Nebula**

Main Stars

STARS	VIEW	COLOR	MAGNITUDE	DISTANCE (LY)
Betelgeuse (alpha)		Red	0.7	425
Rigel (beta)		Blue	0.1	775
Bellatrix (gamma)		Blue-white	1.6	240
Mintaka (delta)		Blue	2.2	915
Alnilam (epsilon)		Blue	1.7	915

THIS MAY be the most recognizable constellation in the sky, visible worldwide from its position straddling the celestial equator, and identified by different cultures across thousands of years. Its prominence earned three references in the

makes the structure of a man in the constellation easy to imagine. The hunter's bright belt is made of three stars all around 2nd magnitude. His upraised, club-wielding arm sweeps into the Milky Way while his other hand holds a lion's belt.

Bright Stars

Orion's main stars are among those that most easily stand out in the night sky. They can provide a way to orient yourself even amid the glare of city lights. Large, reddish tinted Betelgeuse is the alpha star and represents his right shoulder. It is a huge variable star that pulsates to a diameter between 300 and 400 times the size of our Sun. Blue-white Rigel, the hunter's left foot, is a magnitude 0.1 supergiant, 57,000 times as bright as the Sun.

The hunter's left shoulder is formed by another bright star, magnitude 1.6 Bellatrix (and if Orion's cultural centrality were in doubt, the name of this star would clarify it, appearing as an evil character in the Harry Potter series of books). Alnitak, Alnilam, and Mintaka—all around 2nd magnitude or less—form Orion's distinctive diagonal belt, one of the

characteristics by which it is most easily recognized.

Mythology

There are a few stories in Greco-Roman mythology associated with Orion. In one, Orion was a powerful hunter dealt a lethal sting by Scorpius—a reason for the two being placed, as they are, on opposite sides of the sky.

In another story, Artemis, the goddess of the hunt, admired Orion for his great skill as a hunter. Her twin brother, Apollo, grew jealous that his sister was showing favor to another man. He tricked her into shooting Orion with a deadly arrow. Upon discovering what she had done, Artemis immortalized Orion in the

stars. Now, the hunter also engages with the animals close to him, such as Taurus (the Bull) and Lepus (the Hare), along with his faithful hounds, Canis Major and Canis Minor.

INTERESTING OBJECTS in Orion

Orion includes an area of the Milky Way that features intense star production. Beneath the hunter's belt three stars form Orion's sword, and in the middle of them is the GREAT NEBULA (M42). Visible as a cloud where star formation is taking place at a furious pace. At its center are the four stars known as the TRAPEZIUM. Another nebula, M43, lies just a bit to the north. For a challenge, try hunting south of Zeta Orionis for the HORSEHEAD NEBULA, a thick black cloud that takes on a horsehead shape, blotting out light from a nebular cloud behind it.

214

Backyard Guide to the Night Sky

SKY FACT — Orion is among the constellations to work its way into classic texts. It is mentioned three times in the Bible, for example, including this cosmic question from Job: "Can you bind the chains of the Pleiades, Or loose the cords of Orion?"

215

CHAPTER 4 Constellations

❷ **STAR MAPS**

❹ **INTERESTING OBJECTS BOX**

sky charts in Chapter Seven to find the constellation in the larger context of the night sky. "Size on the Sky" uses outstretched hands and closed fists to indicate the constellation's approximate size in the sky if you use your hands to measure it. The constellation's alpha star is noted, as is a key "Deep Sky Object"—something of particular interest that makes a good target for binoculars or a telescope. You will find more of these on the diagrams with their associated M—or Messier—numbers. The few objects without Messier numbers have other labels.

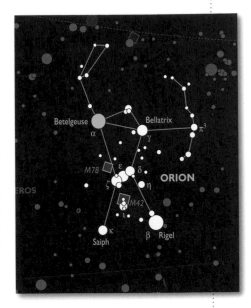

❷ STAR MAPS

Every constellation features a star map, showing the main constellation, the interesting objects within it, and neighboring night sky objects. Light lines connect the member stars of each constellation, while prominent asterisms are also named. The stars down to magnitude 3.5 are labeled following the Bayer system, with the brightest star generally given the Greek letter alpha, the next brightest beta, and so on.

A few of the brighter and more well-known stars will be labeled with their proper names and, where appropriate, tinted to indicate the approximate color of the star in the sky. For fainter constellations, the brightest star will always be labeled. Background stars, neighboring constellations, and nearby deep sky objects are included to help orient your view of the night sky.

❸ MAIN STARS TABLE

For constellations with bright stars, this table lists the most easily seen. The star's name along with it's Bayer designation are included.

❹ INTERESTING OBJECTS BOX

For constellations with notable features and interesting objects, these boxes are filled with more details and facts about them along with advice on the best ways to view them, whether by naked eye, binoculars, or telescope.

When trying to spot constellations, remember to keep this guide handy in the field—and to always use a dim red flashlight when referring to it. White light hurts your night vision and that of your fellow backyard astronomers as well.

Happy hunting . . .

Andromeda: The Chained Maiden

MAKEUP: **7 stars**

BEST VIEWED: **Oct. / Nov.**

LOCATION: **Autumn, center of chart**

SIZE ON THE SKY: 🖐

ALPHA STAR: **Alpheratz**

DEEP SKY OBJECT: **Andromeda galaxy**

I N ANCIENT Greek myth, Princess Andromeda of Ethiopia was chained to a rock as a sacrifice to the gods. In the sky, Andromeda is a V-shaped constellation that is most often seen upside down. To locate the constellation, on fall evenings trace a line northeast from the northeast corner of the Great Square of Pegasus.

The constellation most notably contains the Andromeda galaxy, a spiral-shaped galaxy not unlike our own Milky Way. It is visible on a dark night to the naked eye, a light splotch just west of her right arm.

Mythology

Andromeda's mother, Cassiopeia, claimed that her daughter's beauty surpassed that of the daughters of Nereus, god of the sea and father-in-law of Poseidon. Angered by her boasting, Poseidon sent the monster Cetus to destroy the kingdom of Ethiopia until Cassiopeia and her husband Cepheus sacrificed their daughter. The princess was rescued just in time by Perseus, who was returning home with the head of Medusa, an unlikely weapon that turned Cetus into stone. The whole cast of the tale—parents Cepheus and Cassiopeia, Cetus, hero Perseus, and winged horse Pegasus—are all nearby.

> **SKYFACT**
> The Andromeda galaxy is visible to the naked eye. With binoculars, neighboring galaxies M32 and M110 can also be seen.

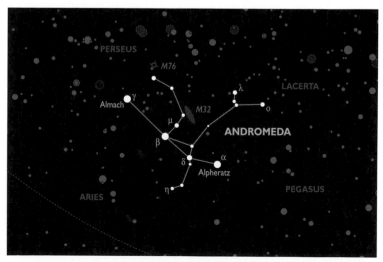

Antlia: The Air Pump

MAKEUP: 4 stars
BEST VIEWED: Mar. / Apr.
LOCATION: Spring, southwest quadrant

SIZE ON THE SKY: 👐
ALPHA STAR: Alpha Antliae
DEEP SKY OBJECT: NGC 2997, spiral galaxy

THERE IS NOT MUCH remarkable about this small constellation, formerly named Antlia Pneumatica. It is primarily a southern constellation that can be picked out in the springtime by stargazers in the Northern Hemisphere when looking toward the southern horizon. Its position lies under the sprawling form of Hydra, about five fist-widths south of the star Regulus in Leo. Antlia's alpha star is its brightest one but has been given no proper name.

Just inside the corner of the constellation, there is a distant spiral galaxy, Antlia's deep sky object, NGC 2997. To an amateur astronomer, this large, faint galaxy is visible—yet difficult to see—through a small telescope. It is very difficult to see with smaller equipment. With much more powerful, professional equipment an astronomer can discern the galaxy's spiral shape, the blue stars populating its arms, clouds of pink hydrogen gas, and a stellar nucleus.

History

Antlia is a fairly recently recognized constellation—most ancient constellations are found in the northern sky, while ones closer to the southern skies are among the newest. Constructed by French astronomer Nicolas-Louis de Lacaille during his trip to the Cape of Good Hope in the early 1750s, it was named to honor the machine invented by Irish chemist, physicist, and inventor Robert Boyle.

Lacaille traveled south to study the skies and stars not visible in the Northern Hemisphere. He spent much of his time observing the skies of the Southern Hemisphere from the Cape of Good Hope, far below the Equator. During his time there, he cataloged over 10,000 new stars and divided previously understudied skies into 14 new constellations, including Antlia. Lacaille's work was published posthumously in 1763; in addition to the thousands of stars and new constellations, he also created a table of eclipses going forward 1,800 years.

INTERESTING OBJECTS in Antlia

ZETA 1 and ZETA 2 ANTLIAE are two notable binary stars visible in the constellation. Zeta 2 is a wide 6th-magnitude double star that you can split with binoculars. Zeta 1 is also a binary star, but you need the power of a telescope to see it as a double.

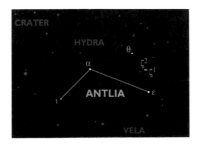

173

Aquarius: The Water Bearer

MAKEUP: **13 stars**

BEST VIEWED: **Sept. / Oct.**

LOCATION: **Autumn, southwest quadrant**

SIZE ON THE SKY: 🖖🖖

ALPHA STAR: **Sadalmelik**

DEEP SKY OBJECT: **M2, star cluster**

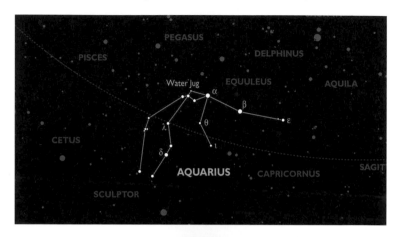

AMONG THE FAINTER constellations of the zodiac, Aquarius lies between Pisces and Capricornus on that great wheel. From the Northern Hemisphere it is found south of the easily recognized square of Pegasus. It is most visible in the fall months, when it reaches its highest point in the middle of the southern sky.

Two planetary nebulae occupy Aquarius. Both are visible by telescope. The Helix Nebula is the closest to Earth, so despite its relatively low density it appears as large as half the diameter of a full Moon. The Saturn Nebula is smaller and more distant, but, when found, it appears as a brilliant green-tinted point.

Mythology

Aquarius represents several mythological characters. The ancient Egyptians deified him for bringing the yearly Nile River flood to their parched croplands: The zodiacal symbol for Aquarius is the hieroglyphic character for water. The constellation is also seen as Ganymede, cupbearer for Zeus. To some, he is responsible for pouring water or wine from a jug, maintaining the mighty celestial river, Eridanus.

SKYFACT

Many of the prominent stars in this constellation have Arabic names suggesting the principles of good fortune. Sadalmelik, the name of the alpha star, translates to "lucky one of the king"—an association consistent with Egyptian visions of this constellation as important to the vital Nile flood.

Aquila: The Eagle

MAKEUP: 10 stars
BEST VIEWED: Aug. / Sept.
LOCATION: Summer, southeast quadrant

SIZE ON THE SKY: 🖐
ALPHA STAR: Altair
DEEP SKY OBJECT: Eta Aquilae, variable star

AQUILA, named by star gazers in ancient Mesopotamia, flies close enough to the Earth's Equator to be seen from anywhere in the world. But it is most easily found in the Northern Hemisphere when looking south in the middle of the summer months.

Altair (Arabic for "the bird"), is the constellation's brightest star and represents the eagle's head. It is also part of the trio of stars that forms the Summer Triangle asterism, and as one of the brightest stars in the sky makes a good reference point for identifying the constellation. The supergiant Eta Aquilae is another notable feature in the constellation. It is a variable star that ranges from magnitude 4.1 to 5.3 every 7.2 days.

Mythology

The eagle belonged to the Greek god Zeus and, according to myth, carried the god's thunderbolts for him. He was also said to have brought to the sky a young shepherd named Ganymede who was to serve as Zeus's cupbearer. The youth would be immortalized as the nearby constellation Aquarius. Aquila has been visualized as flying in two different directions. Arab astronomers referred to the star Zeta Aquilae as Deneb al Okab, the "tail of the eagle." In modern diagrams the star represents one of the bird's wings.

Main Stars

STARS	VIEW	COLOR	MAGNITUDE	DISTANCE (LY)
Altair (alpha)	👁	White	0.8	16.8
Alshain (beta)	👁	White	3.7	45
Tarazed (gamma)	👁	Yellow	2.7	460
Eta Aquilae	👁	Yellow/White	3.9	1,170

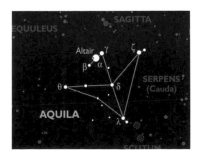

Altair and Tarazed (upper right)

Aries: The Ram

MAKEUP: 4 stars
BEST VIEWED: Nov. / Dec.
LOCATION: Winter, southwest quadrant

SIZE ON THE SKY: 🖐
ALPHA STAR: Hamal
DEEP SKY OBJECT: Gamma Arietis, double star

THE ANCIENT astronomers who organized the zodiac noted that the Sun was "in" Aries at the vernal equinox—meaning it was traveling in the Ram's part of the sky on the day when it passed from the southern to the northern celestial sphere for the year. Precession has since shifted this constellation's position with respect to the Sun's equatorial crossing, but by tradition Aries remains where the zodiac begins. On evenings in late fall and early winter, Aries is high in the east, between the Great Square of Pegasus and the Pleiades in Taurus.

The tail of the Ram is represented by the star Gamma Arietis, a double star with a wide, eight-arc-second separation between the two members. It was one of the first doubles spotted with a telescope—by astronomer Robert Hooke in 1664.

Mythology

Aries was widely recognized as a ram—despite being made of only four stars. According to the Greeks, Aries was the source of the Golden Fleece stolen by Jason and the Argonauts. The ram was sent to rescue two children of a king from an abusive stepmother. When the ram returned, the grateful king sacrificed it and left the fleece in the custody of a dragon—from which it was stolen by Jason.

Main Stars

STARS	VIEW	COLOR	MAGNITUDE	DISTANCE (LY)
Hamal (alpha)	👁	Orange	2.0	66
Sheratan (beta)	👁	White	2.6	60
Mesartim (gamma)	👁	White	4.8	204

Auriga: The Charioteer

MAKEUP: 7 stars
BEST VIEWED: Dec. / Jan.
LOCATION: Winter, center of chart

SIZE ON THE SKY: ✋
ALPHA STAR: Capella
DEEP SKY OBJECT: M36 & M37, star clusters

AURIGA IS an elegant constellation in the heart of the Milky Way. It is easily identified by its alpha star, the brilliant Capella, the sixth brightest star in the sky. When it is at its highest in the early winter months, Auriga can be found on a line between Orion and Polaris. Auriga is an ancient constellation, one of Ptolemy's original 48.

Epsilon Aurigae, the star just southwest of Capella, is an eclipsing binary veiled every 27 years by a companion whose nature is still not fully known. The next eclipse is scheduled to begin in 2009 and will reduce the 3.5-magnitude star by about half a magnitude. The deepest part of the eclipse may last as much as a full year. Some scientists believe this behavior indicates that Epsilon Aurigae's companion is surrounded by clouds of gas and dust.

Deep Sky Objects

Straddling the galactic equator, Auriga provides a window on several interesting star clusters visible through binoculars. One bright highlight is open cluster M36, which can be found just five degrees southwest of Theta Aurigae. Another open star cluster, the heavily populated M37 can be seen with either binoculars and a small telescope. Through binoculars it will appear as a fuzzy spot, while through the telescope a large number of distinct stars can be seen.

Mythology

There are a few stories associated with this constellation. The star Capella represents a mother goat that the charioteer carries on his back along with her three kids (the neighboring stars). Another variation in the constellation's legend is that the chariot and rider may represent Hephaestus, the crippled blacksmith to the gods, who built the vehicle to move about more easily.

SKYFACT
Auriga and Taurus share a star. Beta Tauri would be considered Auriga's gamma star if it were part of the constellation.

CHAPTER 8: Constellations

Boötes: The Herdsman

MAKEUP: 8 stars

BEST VIEWED: June

LOCATION: Summer, center of chart

SIZE ON THE SKY: 🖐

ALPHA STAR: Arcturus

DEEP SKY OBJECT: Izar, double star

THIS ANCIENT constellation is one of the most distinct in the early summer sky. Boötes is most easily identified by its brightest star, Arcturus. It is simple to find this brilliant star not only because it is the fourth brightest in the sky, but because it is on an arc that connects to the handle of the Big Dipper—a reference point for the "arc to Arcturus." Arcturus represents the knee of Boötes. The herdsman's head is pointed toward Polaris as he runs across the night sky with one arm— upside down from a northern perspective.

The Quadrantid meteor shower falls from the northern part of the constellation Boötes and is one of the strongest of the year, producing several dozen meteors an hour. It takes place in the first week of January and originates at the meeting point of Boötes, Hercules, and Draco. Quadrantid meteors, however, are visible in substantial numbers only during a period lasting a few hours.

Mythology

There are many different legends surrounding Boötes. According to some, he is there to keep his flock of creatures moving about the sky, as he pursues the bears, Ursa Major and Ursa Minor. This version is supported by the loose translation of Arcturus as "Bear Guard." In another incarnation he is the son of Callisto—the paramour of Zeus who was changed into a bear, Ursa Major, by Zeus's angered wife. He is also known as the son of Demeter, the goddess of the harvest, granted a place in the stars for inventing the plow.

Main Stars

STARS	VIEW	COLOR	MAGNITUDE	DISTANCE (LY)
Arcturus (alpha)	👁	Yellow/Orange	-0.1	37
Nekkar (beta)	👁	Yellow	3.5	219
Seginus (gamma)	👁	White	3.0	85
Izar (epsilon)	👁	Orange	2.7	300

Camelopardalis: The Giraffe

MAKEUP: 5 stars
BEST VIEWED: Dec. / Jan.
LOCATION: Winter, northwest quadrant

SIZE ON THE SKY: 🖐
ALPHA STAR: Alpha Camelopardalis
DEEP SKY OBJECT: NGC 1502, star cluster

Spiral galaxy NGC 2403

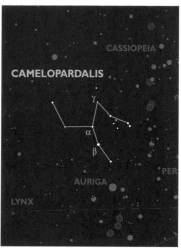

THIS MODERN constellation, created in 1613, filled the open space between the bears, Ursas Major and Minor, and Perseus. It lies right beside Polaris, making it a strictly northern constellation.

In the middle of the constellation, star cluster NGC 1502 is a good telescope target. Stretching toward the cluster is a string of stars known as Kemble's Cascade, while the cluster itself contains two easily split double stars—a 5th-magnitude pair and a 9th-magnitude pair. An irregular variable star, VZ Camelopardalis, lies northwest of the constellation, near Polaris. Its variation is slight and may be difficult for the unpracticed eye to observe.

Another object is NGC 2403, a spiral galaxy that lies about 12 million light-years from Earth. In 2004, a supernova was discovered here.

History

To the ancient Greeks, this group of stars represented a "camel-leopard," an animal with the head of a camel and the spots of a leopard. However, it was not officially named until 1613 by the Dutch astronomer and theologian Peter Plancius. He named the constellation to honor the camel that brought Rebecca to Isaac in the biblical story.

> **SKYFACT**
> The magnitude of Z Camelopardalis, a cataclysmic variable star, ranges from 9.6 to 13. It is visible every night of the year.

179

Cancer: The Crab

MAKEUP: **5 stars**
BEST VIEWED: **Mar. / Apr.**
LOCATION: **Spring, southwest quadrant**

SIZE ON THE SKY: 🖐
ALPHA STAR: **Acubens**
DEEP SKY OBJECT: **Beehive cluster, M44**

TINY CANCER, the Crab, lies between Gemini and Leo on the zodiac. In the sky, it can be found just northeast of Orion's belt. Made of five dim stars, the constellation is best visible in spring during March and April when it is at its highest position in the sky. Despite being a member of the zodiac, Cancer is relatively faint, with no stars brighter than 4th magnitude.

Deep Sky Objects

Cancer includes one of the more interesting groups of stars in the night sky: The Beehive cluster (M44) is located just east of the Crab's head (the star in the center of the constellation). Visible to the naked eye under dark skies, the cluster's more than 20 stars will come into view nicely through binoculars, even under urban skies.

Other deep sky objects are best seen with telescopes. The open cluster M67 has 500 stars. The cluster can be seen with binoculars but the best view comes through a small telescope. The star R Cancri is of interest as well; this long-period variable star's magnitude ranges from 6.2 to 11.2. Its brightness takes approximately one year to go from maximum to minimum. Although it can be seen with binoculars at its brightest, it is easier to track its changes with a telescope.

Mythology

Cancer was given a place in the sky by Hera, queen of the gods, who was attempting to thwart the efforts of Hercules, her husband Zeus's son by a mortal woman. She sent the crab to distract the strong Hercules in his battle with the many-headed monster, the Hydra, one of his famous 12 labors. Hardly a fair fight, the crab was promptly crushed under the warrior's foot—but it did earn a place in the stars.

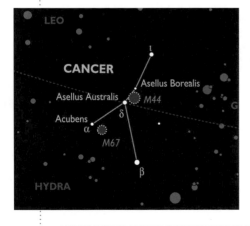

SKYFACT

Early mapmakers identified the Tropic of Cancer as the northernmost latitude at which the Sun appears directly overhead at noon. This occurred when the Sun was in Cancer during the summer solstice. Since then, the Sun's position in the sky on that date has shifted, but the line on our globe is the same.

Canes Venatici: The Hunting Dogs

MAKEUP: 2 stars
BEST VIEWED: May / June
LOCATION: Spring, center of chart

SIZE ON THE SKY: 🖐
ALPHA STAR: Cor Caroli
DEEP SKY OBJECT: Whirlpool galaxy, M51

THIS TWO-STAR constellation can be found just below the handle of the Big Dipper—or, alternatively, look for it running between the legs of Ursa Major, the bear that the dogs are chasing.

The two stars of Canes Venatici are surrounded by a multitude of interesting objects, some visible to the naked eye. Y Canum Venaticorum, or La Superba, is a bright red star of 5th magnitude, which varies over about five months to magnitude 6.6. Cor Caroli—the Heart of Charles—is the bright alpha star of Canes Venatici, reputedly named by Edmond Halley for his sponsor, Charles II.

Deep Sky Objects

A 6-inch-aperture telescope will begin to reveal the spiral structure of the well-known Whirlpool galaxy (M51). This 8th-magnitude object can be found near the border of Canes Venatici on a line traced from Cor Caroli to the final star in the handle of the Big Dipper.

Another site is the globular cluster M3, which can be found midway between Cor Caroli and Arcturus in Boötes. The cluster is made of up to 500,000 stars, is located roughly 34,000 light-years from Earth, and has

CANES VENATICI

a magnitude of 6.2. On clear nights M3 can be seen with the naked eye, and through a small telescope, stars can be defined.

Mythology

Cor Caroli and Chara (beta Canum Venaticorum) represent Asterion and Chara, two leashed hunting dogs. These two hounds are close in the sky to their master Boötes, leading him in his hunt for the bears, Ursa Major and Ursa Minor. This small northern constellation was conceived by Johannes Hevelius in the 17th century and has since been accepted as one of the official 88 constellations.

INTERESTING OBJECTS
in Canes Venatici
SPIRAL GALAXY M51, possibly the first spiral galaxy to be cataloged and certainly one of the most beautiful, is over 60,000 light-years across and about 31 light-years away. Currently, M51 is interacting with the galaxy NGC 5195, a dwarf galaxy whose gravitational disturbance is triggering star formation in the larger galaxy.

181

Canis Major: The Larger Dog

MAKEUP: **8 stars**

BEST VIEWED: **Jan. / Feb.**

LOCATION: **Winter, southeast quadrant**

SIZE ON THE SKY: 🖐

ALPHA STAR: **Sirius**

DEEP SKY OBJECT: **M41, star cluster**

CANIS MAJOR is located near the Milky Way just to the east of Orion. The constellation is most easily spotted by locating Sirius, the brightest star in the night sky. Only 8.6 light-years away, it shines a brilliant magnitude -1.5. Sky watchers in the south can identify Sirius by tracing a line through Orion's belt and continuing southeast, where the bright star marks the northern edge of Canis Major.

The constellation is the location of several clusters and nebulae listed in the Messier and NGC lists. The brightest is open cluster M41, easy to spot with binoculars and impressive in a telescope.

Mythology

Generally Canis Major is considered to be one of Orion's hunting dogs. The phrase "dog days of summer" takes its origin from Sirius. In late summer the star rises around the same time as our Sun, leading to the belief that its heat conspires with the Sun to bring extra warmth to the Northern Hemisphere at that time of year.

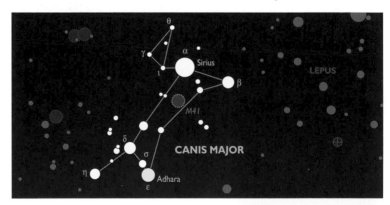

Main Stars

STARS	MAGNITUDE	DISTANCE (LY)
Sirius (alpha)	-1.5	8.7
Murzim (beta)	2.0	500
Muliphein (gamma)	4.1	400
Wezen (delta)	1.8	1/91
Adhara (epsilon)	1.5	430

Sirius outshines all other stars.

Canis Minor: The Smaller Dog

MAKEUP: 2 stars
BEST VIEWED: Jan. / Feb.
LOCATION: Winter, southeast quadrant

SIZE ON THE SKY: 👎
ALPHA STAR: Procyon
DEEP SKY OBJECT: none

ORION'S OTHER hunting dog lies directly northeast of its larger companion, Canis Major. A dimmer and smaller constellation (as the name implies), Canis Minor is not as bright or vibrant as its counterpart. Visualizing a young dog in this small set of stars takes some creativity, but the angle at which Canis Minor lies is evocative of the animal's cocked head as he looks toward his northern neighbor, Gemini.

A challenge for the beginning backyard astronomer, this constellation is composed of only two stars brighter than 5th magnitude and contains no deep sky objects brighter than the 15th magnitude. The alpha star, Procyon, is the brightest in Canis Minor and the eighth brightest star in the night sky. This deep yellow star is only 11.2 light-years away from Earth. In midwinter from the Northern Hemisphere, the star is located in the middle of the sky to the south, and forms a bright triangle with Betelgeuse to the west and Sirius to the south.

A faint meteor shower (magnitude 6 to 11) falls from Canis Minor in early December. The Canis Minorids were discovered by accident while an astronomer was studying another shower in 1964.

Mythology

In addition to being considered one of Orion's hunting dogs, Canis Major has a few other associations as well. Some surmise that he rests under the table expecting scraps from Castor and Pollux, the twins of Gemini. Another myth is that Canis Minor represents Helen of Troy's favorite dog that allowed her to elope with the Trojan prince, Paris. Canis Minor was also said to belong to a mortal hunter, Actaeon, who had the misfortune to gaze upon the goddess Artemis while she was bathing. Enraged, she turned Actaeon into a stag, who was then set upon and eaten by his own dogs.

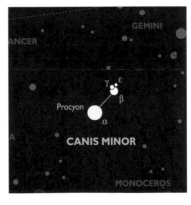

183

SKYFACT Another seasonal signpost: The Winter Triangle asterism is formed by Canis Minor's Procyon, Orion's Betelgeuse, and Canis Major's Sirius. This asterism is the bright, seasonal counterpart to the Summer Triangle asterism formed by Altair, Deneb, and Vega.

Capricornus: The Sea Goat

MAKEUP: **12 stars**

BEST VIEWED: **Aug. / Sept.**

LOCATION: **Summer, southeast quadrant**

SIZE ON THE SKY: ✋

ALPHA STAR: **Algedi**

DEEP SKY OBJECT: **M30, star cluster**

THE STARS forming this zodiacal constellation are not among the brightest—3rd magnitude at the most—but their broad, triangular shape is easily recognizable under a dark sky. For sky watchers in the Northern Hemisphere, Capricornus is most visible during the late summer and early fall in the southern sky. The constellation lies southeast of the bright Altair in Aquila. Capricornus is also bordered by Sagittarius, Piscis Austrinus, and Aquarius.

Visible to the naked eye on a clear night, Algedi, Arabic for "the goat," is Capricornus' alpha star—actually two of them. It is an optical binary, apparent to the naked eye on nights of good seeing, composed of two stars that are closely aligned but separated by more than 500 light-years.

The Tropic of Capricorn represents the latitude where the Sun reaches its winter solstice, the southernmost position in the sky. Early astronomers observed that this happened when the Sun was in front of Capricornus. Due to shifts in the celestial coordinates due to precession, the winter solstice is no longer in Capricornus, but the position on the globe is unchanged.

Mythology

Long recognized as a goat, Capricornus was subsequently given a fish tail. In one story the god Pan leaped into the River Nile to escape a monster, and the water transformed him. In an older tale, the animal represents one of Zeus's warriors in the battle with the Titans. He discovered conch shells, whose resounding call frightened the Titans into retreat. In appreciation, Zeus placed him in the sky with a fish tail and horns to represent his discovery.

See "The Sun's Path," pp. 62-63

184

Backyard Guide to the Night Sky

INTERESTING OBJECTS in Capricornus

ALGEDI, the alpha star in the constellation Capricornus, is only the third brightest star in the constellation. However, it is a naked-eye "double star." The pair is not actually a true double-star pair, but a very impressive illusion, called an optical or line-of-sight double where the two stars are not a true binary system. Alpha 1 is located about 690 light-years away whereas Alpha 2 is about 109 light-years away. Both have evolved and are dying yellow stars.

Cassiopeia: The Queen

MAKEUP: **5 stars**

BEST VIEWED: **Oct. / Nov.**

LOCATION: **Autumn, northeast quadrant**

SIZE ON THE SKY: 🖐

ALPHA STAR: **Shedar**

DEEP SKY OBJECT: **M52, star cluster**

FROM MUCH of the Northern Hemisphere Cassiopeia is visible all year because of her proximity to the north celestial pole. She sits on a throne within a portion of the Milky Way facing Polaris. Beneath her is Andromeda and above is Ursa Minor. Cassiopeia's obvious W shape (or M, depending on the sea-

Deep Sky Objects

One of several star clusters in the constellation, open cluster M52 is the best one for a beginner to view. It is a dense open cluster that is easily visible through a telescope. Locate it by following the path of the southernmost leg of the W. This cluster contains about 100 stars, making it one of the richest in the northern sky.

Mythology

Cassiopeia is one of the key figures in the story of Perseus and Andromeda. Nearby, the other characters from the famous Greek myth accompany the boastful Ethiopian queen. On her throne she sits, located between her husband, Cepheus, and Andromeda, her daughter.

In the story, Cassiopeia begins the drama by boasting that her daughter's beauty was greater than that of the Nereids, the daughters of the sea god Nereus. In return for offending the gods of the sea, Cassiopeia was forced to sacrifice her daughter's life or else the entire kingdom would be destroyed by a sea monster.

Andromeda escaped her fate when Zeus's son, Perseus, rescued her, but Cassiopeia did not avoid punishment from the gods. Upon her death, the queen was banished to the sky where she suffers chained to her throne, forced to hang upside down for half the year.

Main Stars

STARS	MAGNITUDE	DISTANCE (LY)
Shedar (alpha)	2.2	230
Caph (beta)	2.3	54
Tsih (gamma)	2.5	610
Ruchbah (delta)	2.7	~100

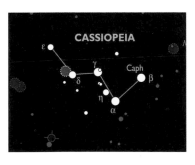

CASSIOPEIA

son) makes this constellation one of the easiest to identify. Although she is visible year-round, the best time to observe Cassiopeia is in the fall from October through November, when she reaches her highest point.

Cepheus: The King

MAKEUP: **10 stars**

BEST VIEWED: **Sept. / Oct.**

LOCATION: **Autumn, center of chart**

SIZE ON THE SKY: 🖖

ALPHA STAR: **Alderamin**

DEEP SKY OBJECT: **Delta Cephei, variable star**

A CIRCUMPOLAR constellation, Cepheus is visible all year in the Northern Hemisphere. The constellation is in a relatively empty patch of sky, but can be tricky to pick out because its stars are of a brightness similar to those surrounding it. Five bright stars make up the body of the king. They are in the shape of a small (upside down) house with the roof pointing approximately toward Polaris. Cepheus is also facing the open end in the W of Cassiopeia.

Cepheus played an important part in the discovery of Cepheid variables. The change in Delta Cephei's magnitude was discovered by astronomer John Goodricke in 1784. This star has about a 1-magnitude change of brightness, from 3.5 to 4.4, over about five and a half days. Cepheids with longer periods of variation have greater intrinsic brightnesses. Thus when astronomers use a Cepheid's period to determine its real brightness and compare this with its apparent brightness, the Cepheid can help astronomers determine how far away the star lies. This relationship yields the distance of deep sky objects that contain this common type of variable star.

Another object of interest in the constellation is Mu Cephei—astronomer William Herschel's "garnet star." This target is a red star with an irregular period of variability.

Mythology

Cepheus was the king of Ethiopia and a luckless character in the myth of Perseus and Andromeda. His queen vainly boasted that their daughter was more beautiful than the daughters of Nereus, causing the enraged Poseidon to release a sea monster. An oracle told the king that he had to sacrifice his daughter Andromeda to appease the offended gods.

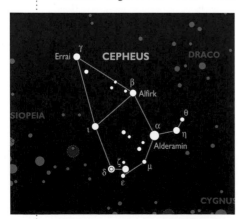

Cetus: The Sea Monster

MAKEUP: **13 stars**
BEST VIEWED: **Nov.**
LOCATION: **Autumn, southeast quadrant**

SIZE ON THE SKY: 🖐🖐
ALPHA STAR: **Menkar**
DEEP SKY OBJECT: **Mira, variable star**

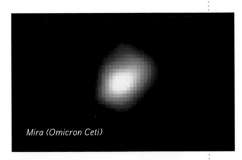

THIS VERY large, faint constellation has a distinctive shape. The area of sky surrounding Cetus is known as the Heavenly Waters and includes Eridanus and Pisces. The constellation is most easily found in autumn in the southern sky with its head between Taurus and Pisces and its body bordering Aquarius. Its head forms a small group of stars connected to the body by Mira, a long-period variable star.

Watching Mira can be a worthwhile project and a helpful way to develop your own sense of stellar magnitude. At its brightest, the star reaches around magnitude 2.4, but then over 11 months it fades to a magnitude of 9.3—invisible to the naked eye.

Mythology

Sea god Poseidon sent the Cetus to terrorize Ethiopia after being insulted by Queen Cassiopeia's boasting that her daughter was more beautiful than the god's wife. The constellation is also said to represent the whale that swallowed Jonah in the Old Testament.

> **SKYFACT**
> The model for Cetus was reportedly a 40-foot (12-meter) skeleton of a sea creature brought to ancient Rome.

Mira (Omicron Ceti)

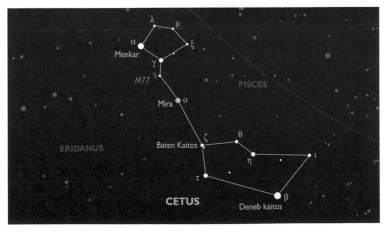

Columba: The Dove

MAKEUP: **8 stars**

BEST VIEWED: **Jan. / Feb.**

LOCATION: **Winter, southeast quadrant**

SIZE ON THE SKY:

ALPHA STAR: **Phact**

DEEP SKY OBJECT: **NGC 1851, star cluster**

THE DOVE is a small-ish constellation, one of the modern ones added in the 16th century to fill out the sky. Columba was the creation of the Dutch theologian and mapmaker, Petrus Plancius. The constellation is best seen in mid-winter during the months of January and February.

The constellation honors the dove that acted as a scout during Noah's voyage on the ark. Besides lying just south of the brilliant star Sirius in Canis Major, Columba's distinguishing feature is that it seems to be three legs spiraling from a central star of medium brightness. When viewing, keep in mind that Columba may be difficult to distinguish for the first-time sky watcher. It is located so close to other night sky objects that are much brighter and may outshine the stars of the Dove.

Deep Sky Object

There is a 7th-magnitude globular cluster to be found in this constellation. To locate it, follow a line traced from Sirius through Wazn (Beta Columbae). This object is NGC 1851. It is easily visible on a clear night through binoculars, while a 6-inch-aperture telescope will begin to reveal the cluster's many individual stars.

History

In the biblical story of the great flood and Noah's ark, Noah released a little dove to search for dry land when the 40 days and nights of rain had ceased. When Columba returned carrying an olive sprig, the good news was spread that the waters had receded and it was safe to return to land. Though Columba is a bird and technically sky bound, its role in the story of Noah's ark makes it a proper member of the Heavenly Waters group, along with the nearby watery constellations Argo, Eridanus, Dolphinus, Pyxis, and Piscis Austrinus.

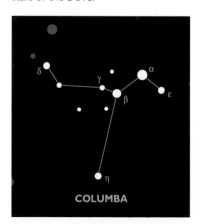

COLUMBA

Main Stars

STARS	MAGNITUDE	DISTANCE (LY)
Phact (alpha)	2.6	270
Wazn (beta)	3.1	86
Ghusn al Zaitun (delta)	3.9	237

Coma Berenices: *Berenice's Hair*

MAKEUP: **3 stars**

BEST VIEWED: **May / June**

LOCATION: **Spring, southeast quadrant**

SIZE ON THE SKY:

ALPHA STAR: **Alpha Comae Berenices**

DEEP SKY OBJECT: **M64, Black Eye galaxy**

COMA BERENICES was once part of nearby constellations—tucked into Virgo as a wisp of her hair, or considered a tuft of Leo's tail. In the 1500s, cartographers and astronomers, including Caspar

Coma Berenices is located just north of Virgo and south of Canes Venatici and the Big Dipper. The proximity to the Virgo cluster of galaxies makes for interesting sky watching. The Black Eye galaxy is visible by telescope in between the two outermost stars in Coma Berenices—effectively at the base of a triangle. With a 4- to 6-inch-aperture telescope, a dark cloud of dust creates the illusion of a "black eye" in the middle of this spiral galaxy.

INTERESTING OBJECTS
in Coma Berenices

BLACK EYE GALAXY (M64) is a spiral galaxy discovered by Edward Pigott in March 1779 and spotted by Charles Messier in 1780. Popular with amateur astronomers because it is visible through small telescopes, the galaxy sports striking dark outer bands that circle its bright nucleus.

Mythology

Berenice was a queen of Egypt during the reign of Ptolemy III. As he went to war, his wife struck a deal with the goddess Aphrodite, promising her long, beautiful hair in return for the safe return of her husband. An astronomer of the royal court convinced the ruling couple that a grateful Aphrodite had placed the queen's gift in the stars.

Vopel and Tycho Brahe, separated this small but rich area of the sky from the others and rejuvenated the tale of an Egyptian queen to name it.

Black Eye galaxy

MAKEUP: 7 stars

BEST VIEWED: July / Aug.

LOCATION: Summer, southeast quadrant

SIZE ON THE SKY: ☝

ALPHA STAR: Alpha Coronae Australis

DEEP SKY OBJECT: NGC 6541, star cluster

Corona Australis Nebula

THOUGH INCLUDED as one of Ptolemy's original 48 constellations, this is primarily a southern constellation. From the mid-northern latitudes it rises only a few degrees above the horizon during the midsummer months. Even then, the Southern Crown is still a rather faint constellation with no star brighter than 4th magnitude. The crown can be found at the feet of Sagittarius and just west of the bright star Shaula at the tail of Scorpius.

Though small, the Southern Crown has been found to host an active star-forming region: The Coronet cluster and Corona Australis Nebula are invisible to all but the most advanced deep sky imaging technology, but photographs show around 30 "newborn" stars. Near the constellation's southern tip, NGC 6541 is a 6th-magnitude globular cluster visible through a small telescope.

Mythology

For a small constellation there are a surprising number of associated myths, most referring to a crown of laurel or fig leaves. Some believe that the crown belonged to Chiron, the Centaur. In another story, Apollo made the crown from the leaves of his transformed love Daphne, who was changed into a laurel tree to escape Apollo's advances.

Main Stars

STARS	MAGNITUDE	DISTANCE (LY)
Alpha CrA	4.1	130
Beta CrA	4.1	510
Epsilon CrA	4.8	98

Corona Borealis: The Northern Crown

MAKEUP: 7 stars

BEST VIEWED: June / July

LOCATION: Summer, center of chart

SIZE ON THE SKY: 🖐

ALPHA STAR: Alphecca

DEEP SKY OBJECT: T Coronae Borealis

AN OBVIOUS crown-like semicircle shape compensates for the small size and faint magnitude of this constellation. Squeezed between Boötes and Hercules, it may be overshadowed by its bigger, brighter neighbors.

You can locate the constellation just south of a line traced between the bright stars Vega in Lyra and Arcturus in Boötes. Corona Borealis is best observed in the early summer months of June and July when its position is highest in the sky.

One of the constellation's stars varies greatly in its brightness, which has been noted on several occasions in different centuries. In 1866 and again in 1946, T Coronae Borealis rose from magnitude 10 to magnitude 2 and then faded back—the result of a nova. When this eruption inevitably occurs again, as is expected, this currently invisible, to the naked eye, star (it is currently about a 10th-magnitude star) will become the brightest one in the constellation.

Mythology

The Northern Crown is primarily associated with another Greek myth and legend. The Greek god of wine, Dionysus, threw his royal crown into the sky to honor and impress Ariadne, his future wife and the princess of Crete. He had proposed to her in the form of a young man, and Ariadne refused him, not wanting to marry someone she thought a mere mortal. His efforts to prove himself a god won her over and the two were married, granting the princess immortality as well. This group of stars is classically considered a crown, but Native American stargazers saw this star formation as a camp circle.

INTERESTING OBJECTS in Corona Borealis

RHO CORONAE BOREALIS is under investigation for extrasolar planets. One planet larger than Jupiter has already been found, and astronomers suspect that it's not alone. At random intervals, this variable star's magnitude drops sharply, ranging from 5.8 to 14.8.

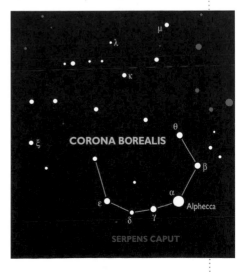

Corvus: *The Crow*

MAKEUP: 5 stars
BEST VIEWED: Apr. / May
LOCATION: Spring, southeast quadrant

SIZE ON THE SKY: ☁
ALPHA STAR: Alchiba
DEEP SKY OBJECT: Ring-tailed galaxy

CORVUS and Crater are both associated in myth with the giant serpent nearby, Hydra. The crow is located near Hydra's tail, and just west of Spica in nearby Virgo. Alchiba, the constellation's alpha star, is just outside the polygon of the bird's body, representing its downward facing head.

The Ring-tailed galaxy is a pair of colliding galaxies, NGC 4038 and NGC 4039 found just outside Corvus. The pair can be seen with an 8-inch-aperture telescope, with the "tail" near the border of Corvus and Crater.

Mythology

The Greek god Apollo sent a crow for a cup of water. While on the errand, the bird was distracted by a tasty-looking unripe fig and stopped to wait for it to ripen. Realizing the delay would anger the god, the crow made up a tale about being attacked by a snake—which he brought back along with the cup of water. Apollo did not stand for his deceit and banished cup, crow, and serpent to the sky.

NGC 4038 and 4039

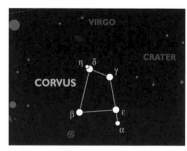

Main Stars

STARS	VIEW	COLOR	MAGNITUDE	DISTANCE (LY)
Alchiba (alpha)	👁	White	4.0	48
Kraz (beta)	👁	Blue-white	2.7	140
Gienah Corvi (gamma)	👁	Blue-white	2.6	165
Algorab (delta)	👁	Orange/White	3.0	88
Minkar (epsilon)	👁	Yellow	3.0	300

Crater: The Cup

MAKEUP: 8 stars
BEST VIEWED: Apr. / May
LOCATION: Spring, southeast quadrant

SIZE ON THE SKY: (☝)
ALPHA STAR: Alkes
DEEP SKY OBJECT: NGC 3887, galaxy

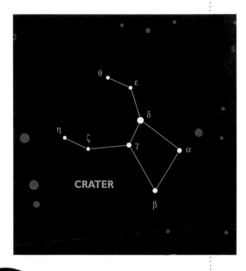

THIS IS A SMALL constellation with numerous associated stories. To some, Crater first represented a spike on the back of the sea monster Hydra. Keeping that in mind will help to more easily locate this dim constellation. From the Northern Hemisphere, Crater appears deep in the southern sky during the mid- to late spring, right above Hydra and just west of Corvus. Locating the star Spica in Virgo may also help as Crater and Corvus are southwest of the bright star. The shape of the cup is unmistakable, four stars establishing the base and four more in a goblet opening toward Spica. There are few bright deep sky objects in Crater; there are some dim galaxies of magnitude 11 or less.

to the constellation Aquarius, the Water Bearer. More recent astronomers have likened the cup to the Holy Grail or to Noah's wine goblet.

Mythology

Most observers of this constellation saw it as a cup. In Greek mythology Crater represents the vessel brought to Apollo by Corvus, the neighboring constellation. The two were tossed to the sky along with Hydra when Apollo realized that Corvus had lied about the reason he was late bringing him a cup of water. Some also relate the cup

Main Stars

STARS	MAGNITUDE	DISTANCE (LY)
Alkes (alpha)	4.0	174
Al Sharas (beta)	4.5	266
Gamma Crateris	4.0	84

193

SKYFACT

Star names in this constellation demonstrate the variety of images in Crater. The name of the alpha star, Alkes, translates as "the cup." The beta star, Al Sharas, translates as "the rib," in reference to its earlier designation as part of Hydra.

Cygnus: The Swan

MAKEUP: 13 stars
BEST VIEWED: Aug. / Sept.
LOCATION: Summer, northeast quadrant

SIZE ON THE SKY: 🖐
ALPHA STAR: Deneb
DEEP SKY OBJECT: North American Nebula

CYGNUS LIES in what for observers is a dense and fascinating portion of the sky. The bird's wings span the Milky Way at a location packed with stars and an assortment of deep sky objects. Its shape and bright stars give it the alternative designation of the Northern Cross—the northern parallel to the brilliant Southern Cross of the Southern Hemisphere. Late summer and early fall see Cygnus highest in the sky with its head pointing south like a bird on its autumn migratory path.

Highlights

Deneb, the alpha star and tail of the bird, is a brilliant first magnitude star that joins Altair and Vega to form the Summer Triangle. The North American Nebula is a large diffuse emission nebula, a colorful gas cloud on the Swan's southern border. It is visible in binoculars under excellent conditions, and a cloud of stars across its face is faintly visible to the naked eye. The nebula's signature resemblance to the North American continent appears most notably in photographs.

Invisible except to a powerful telescope, the Fireworks galaxy (NGC 6946) on the northern border with Cepheus is a spiral galaxy sparkling with new star formation. It is the object of much deep sky photography.

Mythology

This group of stars has been considered a bird for millennia. In Greco-Roman mythology, Zeus transformed himself into a swan to seduce the maiden Leda, who later gave birth to the Gemini twins, Castor and Pollux. The mythical musician Orpheus, son of Apollo, was also thought to be granted this place in the sky. After being murdered for

SKYFACT

A thick band of dust lies within Cygnus creating a dark space in the Milky Way, easily visible to the naked eye under good conditions. This part of the Milky Way is known as the Cygnus Rift and also the Northern Coalsack, a name borrowed from a similar dark spot found in the Southern Cross.

Main Stars

STARS	VIEW	COLOR	MAGNITUDE	DISTANCE (LY)
Deneb (alpha)		White	1.3	2,600
Albireo (beta)		Blue-yellow	3.3	380
Sadr (gamma)		Yellow-white	2.2	1,500
Ruc (delta)		Blue-white	2.9	171
Gienah (epsilon)		Yellow-orange	2.5	72

North American Nebula

scorning the love of a group of young maidens, he was transformed into the shape of a swan to be next to his beloved lyre—the nearby constellation Lyra. The constellation is also considered one of the Stymphalian birds, the hunting of which was among the 12 labors of Hercules.

INTERESTING OBJECTS in Cygnus

Cygnus's part of the sky is worth learning well, with many visible objects and some others that are significant in the history of astronomy. Straddling the galactic equator, Cygnus's beta star is ALBIREO, a bright blue-and-gold binary whose members can be split with even a small telescope. THE VEIL NEBULA, near Epsilon Cygni, is a gas cloud left over from a supernova explosion. The nebula is best viewed using a high contrast filter. Two other objects of note won't be seen through your telescope: CYGNUS A was one of the first galaxies identified by its radio emissions, rather than by light; CYGNUS X-1 was in the mid-1970s conjectured to be the first discovered black hole on the basis of x-rays emitted by gas streaming from a nearby star toward a then unidentified object.

Delphinus: The Dolphin

MAKEUP: 5 stars
BEST VIEWED: Aug. / Sept.
LOCATION: Summer, northeast quadrant

SIZE ON THE SKY: 🖐
ALPHA STAR: Sualocin
DEEP SKY OBJECT: Gamma Delphini

DELPHINUS makes up for its humble size (and dim stars) with a distinct shape. Four stars forming the asterism Job's Coffin make up the body of the animal while one more represents the curved tail.

One of Ptolemy's original 48 constellations, it is still recognized today by the International Astronomical Union. The small constellation seems to be swimming toward Pegasus as it moves through the sky. Delphinus is one of the constellations in the Heavenly Waters, which includes other sea creatures, the mighty ship *Argo,* and the river Eridanus.

In the Northern Hemisphere, Delphinus is most easily seen in August and September when it is overhead. It can be found just west of a straight line traced between Altair in the constellation Aquila, and Deneb in the constellation Cygnus.

Stars & Objects

In 1814 the alpha and beta stars in the group that make up the body of Delphinus were respectively named Sualocin and Rotanev in a star catalog published at the Palermo Observatory. When reversed, the letters in the star names spell "Nicolaus Venator," the Latin form of Niccolo Cacciatore, who was the assistant director of the observatory at the time.

Delphinus's deep sky object is one of its stars, Gamma Delphini. About 101 light-years away, this optical double has a separation of ten arc seconds. The pair is best seen through the view of a telescope. The dimmer star has a slight green tinge.

Mythology

Delphinus was granted a place in the sky because he was a favorite of the Greek sea god Poseidon. The little dolphin is fabled to have convinced the Nereid Amphitrite to marry the god, who had been otherwise unsuccessful at attracting her attention.

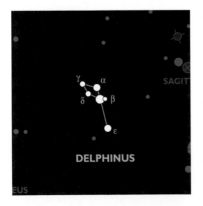

DELPHINUS

Main Stars

STARS	MAGNITUDE	DISTANCE (LY)
Sualocin (alpha)	3.8	240
Rotanev (beta)	3.6	97
Gamma Delphini	5.1	101

Draco: The Dragon

MAKEUP: 18 stars
BEST VIEWED: May / June
LOCATION: Spring, northeast quadrant

SIZE ON THE SKY: 🖐🖐
ALPHA STAR: Thuban
DEEP SKY OBJECT: NGC 6543, planetary nebula

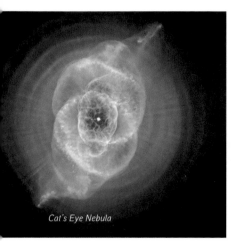

Cat's Eye Nebula

THIS IS ONE of the closest constellations to the north celestial pole, a giant circumpolar shape visible all year in the Northern Hemisphere. Draco's alpha star, Thuban, was once Earth's polestar. Precession has since shifted the celestial position of Earth's axis, and Polaris has taken Thuban's place.

INTERESTING OBJECTS in Draco

CAT'S EYE NEBULA (NGC 6543) is located in Draco. Combined images from the Hubble Space Telescope and the Chandra Observatory show this dying Sun-like star is emitting extremely hot gasses. Scientists predict that the Sun will go thorough a similar process in five billion years.

Although visible year-round, Draco is easiest to spot in the late spring. The dragon's tail begins between Ursa Minor and Ursa Major, dips down toward Cepheus, the King, and ends with its head pointing toward Hercules. The Quadrantid meteor shower erupts from Draco in the beginning of January. It is one of the heaviest meteor showers of the year but lasts only a few hours.

Mythology

Draco has represented various scaly beasts to many different civilizations. Ancient Greeks considered the constellation to be the dragon Ladon, which was slain by Hercules. Hindu mythology sees an alligator in this constellation, while the Persians claimed it to be a giant serpent.

197

Equuleus: The Little Horse

MAKEUP: **4 stars**

BEST VIEWED: **July / Aug.**

LOCATION: **Summer, southeast quadrant**

SIZE ON THE SKY: 🤏

ALPHA STAR: **Kitalpha**

DEEP SKY OBJECT: **None**

EQUULEUS is the second smallest constellation, next to Crux—at least as measured by the area contained within the borders set by the International Astronomical Union. Tiny Canis Minor has fewer stars—only two—but is given a larger patch of space in the sky.

Located in the crowded southern portion of the sky, it is easy to overlook the Little Horse, whose brightest star is magnitude 3.9. It rests between the much larger constellations Pegasus—its equine counterpart, to the northeast—and Aquila, the Eagle, directly southwest. It's neighbors are good "skymarks" to help locate it. July and August provide the best times to view this tiny constellation.

History

Though one of the more inconsequential shapes in the sky, Equuleus' creation is attributed to one of astronomy's early greats—the Greek observer Hipparchus. Still, Ptolemy recognized the constellation as only a partial form of a horse, while Arab astronomers focused on the star Alpha Equulei—also known as Kitalpha—as the only significant feature of the constellation. Some observers even regarded the constellation as subordinate to Pegasus, referring to it as Equus Prior, in reference to the fact that its rising and setting precedes that of the larger constellation.

Mythology

Equuleus is thought to represent Celeris, brother-horse of Pegasus in Greek mythology. While Pegasus is widely known as the winged horse belonging to Perseus, Celeris is said to have belonged to Castor, twin brother of Pollux and also sibling to Helen of Troy (Castor and his twin are the two figures represented in the constellation Gemini). By legend, Castor was a skilled equestrian, and the horse was a gift from the messenger god Hermes.

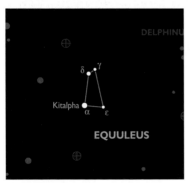

EQUULEUS

SKYFACT

It is sometimes difficult to imagine what people were picturing when they made up constellations. Equuleus, for example, looks nothing like a horse—at least no more, perhaps, than the elongated head of one. As in all constellation-watching, bring your imagination.

Eridanus: The River

MAKEUP: 33 stars
BEST VIEWED: Dec. / Jan.
LOCATION: Winter, southwest quadrant

SIZE ON THE SKY: 🖐🖐
ALPHA STAR: Achernar
DEEP SKY OBJECT: NGC 1300, spiral galaxy

STARGAZERS in the Northern Hemisphere can follow the course of the constellation from its source just west of the foot of Orion until—as will be the case for many observers—it cascades right over the horizon. Its bright alpha star, Achernar, lies at the southern tip of the constellation, but is not visible above roughly 20 degrees north latitude.

Eridanus is the sixth largest constellation in the sky, and Achernar is the sixth brightest star. With an open horizon, much of this massive constellation is visible, especially in the winter. Its first few bends can be seen starting from Rigel, the bright star at Orion's foot. A star very similar to our Sun is being investigated for extraterrestrial life: Epsilon Eridani, also called Sadira, is in the northern bend of the river and can be seen unaided.

Mythology

The constellation has been identified as the Euphrates or the Nile. The ancient Greeks believed it was the river into which Phaethon was cast for being foolhardy enough to believe that he could control the sun god's chariot.

Main Stars

STARS	MAGNITUDE	DISTANCE (LY)
Achernar (alpha)	0.5	144
Cursa (beta)	2.8	89
Sadira (epsilon)	3.7	10

Barred spiral galaxy NGC 1300

Gemini: *The Twins*

MAKEUP: **13 stars**

BEST VIEWED: **Feb. / Mar.**

LOCATION: **Winter, southeast quadrant**

SIZE ON THE SKY:

ALPHA STAR: **Castor**

DEEP SKY OBJECT: **NGC 2392, Clownface Nebula**

M35 open star cluster

OVERHEAD in the late winter, sky watchers can see a group of stars long identified as twins—the famous brothers Castor and Pollux. The feet of the twins of Gemini are just northeast of Betelgeuse, the bright star at the shoulder of Orion's upraised arm. At the heads of the twins, Castor and Pollux are the alpha and beta stars in the constellation.

The annual Geminids is one of the more impressive meteor showers. Look toward Gemini around December 14, when the constellation is in the southeast. An interesting feature in Gemini is only visible by telescope, but noted for its whimsical name. The Clownface Nebula reveals a disk-shaped blue-green planetary nebula. Photographs through large scopes reveal the source of the name—a ring of light around a central star alternately seen as the nose of the clown.

INTERESTING OBJECTS in Gemini

MESSIER 35 resides near the three "foot stars" of Gemini. This open star cluster consists of hundreds of stars. It is about 20 light-years across and about 2,700 light-years distant. Best views of this cluster are in binoculars or a telescope.

Mythology

The alpha and beta stars are named after the twins born of Leda and Zeus. They served as shipmates with Jason on the *Argo*. Castor and Pollux also had a sister named Helen, whose beauty instigated the Trojan War.

Backyard Guide to the Night Sky

Grus: The Crane

MAKEUP: 11 stars
BEST VIEWED: Sept.
LOCATION: Autumn, southwest quadrant

SIZE ON THE SKY: 🖐
ALPHA STAR: Alnair
DEEP SKY OBJECT: Grus Quartet of galaxies

GRUS IS invisible for much of the year in the Northern Hemisphere, making it an almost wholly southern constellation. The crane does emerge toward the south for a short time in the fall. It can be identified by its "X" shape with a bright central star just south of Fomalhaut in Piscis Austrinus.

The three brightest stars in the constellation are unremarkable, but show how apparent magnitude can be affected by distance. Alpha Gruis, also known as Alnair, meaning the "bright one of the tail" in Arabic, is classified as the brightest star partly because it is just 100 light-years away. Alpha Gruis is a large blue star. The second brightest star, Beta Gruis, is more than ten times brighter in an absolute sense, but ranks second because its appearance is dimmed by its distance from Earth of 170 light-years. Gamma Gruis is brighter intrinsically than both stars, but is 200 light-years away.

Within the Crane is the Grus Quartet, a group of four spiral galaxies, NGC 7552, 7582, 7590, and 7599. Discovered in 1827 by Scottish astronomer James Dunlop while he working in New South Wales, Australia, these galaxies are located very close together and exert a strong influence on each other. High starburst activity has been observed between NGC 7552 and 7582. It is believed that the four galaxies will eventually unite, although the process will take millions of years.

Main Stars

STARS	MAGNITUDE	DISTANCE (LY)
Alnair	1.7	100
Beta Gruis	2.1	170
Gamma Gruis	3.0	200

History

The modern designation of Grus as a crane is 400 years old. Until the 1600s, Grus was considered part of Piscis Austrinus. The constellation was broken out and named in 1603 by Johann Bayer in honor of the bird whose high flight led the ancient Egyptians to make it the symbol of astronomy. Some alternate names for the star group include Anastomus (the Stork), Den Reygher (the Heron), and Phoenicopterus (the Flamingo).

Hercules

MAKEUP: 20 stars
BEST VIEWED: July / Aug.
LOCATION: Summer, center of chart

SIZE ON THE SKY:
ALPHA STAR: Rasalgethi
DEEP SKY OBJECT: M13, Hercules Cluster

ONE OF THE MORE prominent heroes of Greek mythology, Hercules has earned a central place in the sky as well—at least for those in the Northern Hemisphere. With a group of central stars that form a recognizable "back-

> **SKYFACT**
> Five other constellations have a mythological connection to Hercules—Cancer, Hydra, Cygnus, Sagitta, and Sagittarius.

wards K," the constellation will appear almost directly overhead during the northern summer months.

From mid-northern latitudes the warrior can be seen running east toward the Milky Way. His right arm is upraised holding a club and both knees are bent. Four stars in an approximate square form his torso, and also make up the asterism known as the Keystone, similar in shape to the central stone in archways. Right next to bright Vega, Hercules is easy to locate, especially during the summer weeks when the constellation is at its highest.

Though prominently positioned in the night sky, Hercules does not benefit from any notably bright stars. Only three are brighter than 3rd magnitude. Rasalgethi, the alpha star, is a red star that is visible to the naked eye. Despite his lack of bright stars, Hercules has nevertheless been a commonly accepted shape in the sky.

Mythology

In Greco-Roman lore, Hercules, a half-mortal son of Zeus known for his amazing strength, had been driven mad by Hera, queen of the gods and Zeus's wife. While out of his mind, Hercules killed his wife and children. When sanity returned, Hercules was overcome with grief. He undertook 12 labors to repent for his actions. Each challenge seemed impossible, but through strength and ingenuity, Hercules overcame each one and was granted celestial immortality.

Main Stars

STARS	VIEW	COLOR	MAGNITUDE	DISTANCE (LY)
Rasalgethi (alpha)		Red	3.5	380
Kornephoros (beta)		Yellow	2.8	148
Biceps (gamma)		Red	3.7	195
Sarin (delta)		White	3.1	79
Marsic (kappa)		Light yellow / Pale red	4.7	367

INTERESTING OBJECTS in Hercules

Considered the best globular cluster visible in the northern sky, the HERCULES CLUSTER (M13) is found on the western side of the hero's torso, on one edge of the Keystone asterism. A slight blur to the naked eye, a 6-inch-aperture telescope will begin to show its individual stars, while larger equipment will reveal its bright and densely packed core. A second cluster, M92, lies northeast from M13. As an exercise in distinguishing the colors of stars in a telescope, try splitting the constellation's alpha star, RASALGETHI, a red supergiant, from its blue-green binary companion. Almost directly west of Vega, Rasalgethi is Arabic for "the kneeler's knee"—though it is actually at the foot of Hercules in modern diagrams.

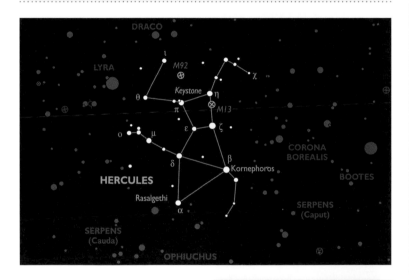

Other cultures assigned figures of great prominence to this group of stars as well. Babylonian astronomers depicted this group of stars as Gilgamesh, the mythical king of ancient Mesopotamia, and hero of one of the earliest known works of literature. Phoenicians along the Mediterranean coast associated the stars with their god Melkarth—which remains the name of the one of the stars in the constellation.

Hercules cluster

Hydra: *The Sea Serpent*

MAKEUP: **17 stars**

BEST VIEWED: **Mar. / Apr.**

LOCATION: **Spring, southwest quadrant**

SIZE ON THE SKY:

ALPHA STAR: **Alphard**

DEEP SKY OBJECT: **M83, spiral galaxy**

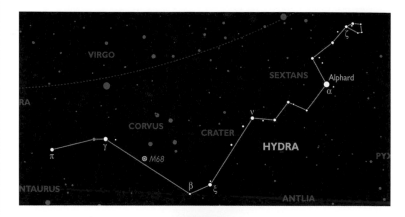

THIS MASSIVE monster occupies a larger strip of sky than any other constellation as it winds its way from Libra to Cancer. Its line of stars is led by a kite-shaped cluster, representing the serpent's head. Spring provides the clearest view. Hydra used to include today's constellations Sextans, Crater, and Corvus.

Alphard, the "heart of the serpent," is Hydra's brightest star and can be located by looking just east of a line traced from Regulus in Leo to Sirius in Canis Major. With a tail that drops close to the southern horizon from mid-northern latitudes, Hydra does not match its size with luminosity: Save for two, its dispersed members are all magnitude 3 or dimmer.

Mythology

When Hydra battled with Hercules, the multiheaded serpent appeared invincible when, as Hercules chopped off each head, more grew in its place. Finally, Hydra was defeated when each stump was burned to prevent the growth of new heads.

INTERESTING OBJECTS in Hydra

The notable objects in Hydra meander across a long stretch of sky. The Mira star R HYDRAE has been recognized as a variable since the late 17th century. Watch it over a year and the luminosity will very from a naked-eye 3.5 to a dim 10.9. V HYDRAE is worth hunting for its color—a lush red—but the real oddity is its composition: Fusion within the star is producing carbon, rather than helium. M83 is a spiral galaxy of the 8th magnitude, and a good hunting ground for supernovae.

Lacerta: The Lizard

MAKEUP: 8 stars

BEST VIEWED: Oct.

LOCATION: Autumn, center of the chart

SIZE ON THE SKY: ✋

ALPHA STAR: Alpha Lacertae

DEEP SKY OBJECT: BL Lacertae

ANOTHER recent addition to the catalog of starry constellations, this northern constellation was identified and named by Johannes Hevelius. Lacerta was first a small mammal, but soon developed into a lizard.

It is a small group of fairly faint stars, but the zigzag shape is easy to recognize on a dark night. It forms a W, similar to the shape of neighboring Cassiopeia. Lacerta can be found between Cygnus, Cassiopeia, and Andromeda. The brighter stars of its neighbors can overshadow Lacerta, making observation a challenge for a first-time viewer.

In late September and early October the constellation will be almost directly overhead for observers in the mid-northern latitudes. That's the best time for a glimpse of Lacerta, though its position makes it visible for much of the year.

Stars & Objects

None of the stars in this constellation have merited proper names, and the most distinct object, BL Lacertae, is only within reach of the most powerful amateur telescopes. But it is worth noting. The center of an elliptical galaxy, it varies in magnitude between 13 and 16.1, and may have at its core a black hole. Lacerta does not contain any Messier objects, no galaxies brighter than magnitude 14.5, and no star clusters.

History

Because this constellation was so recently created—filling one of the gaps in the sky left by the ancient designations—there is no relevant mythology or associated backstory. Indeed there were competing efforts to name this piece of stellar turf in honor of Louis XIV and Frederick the Great.

Main Stars

STARS	MAGNITUDE	DISTANCE (LY)
Alpha Lacertae	3.8	102
Beta Lacertae	4.4	170
10 Lacertae	4.9	1,000 - 1,060

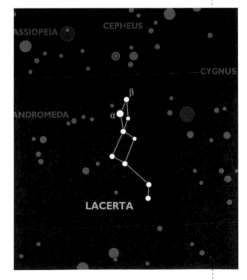

Leo: The Lion

MAKEUP: **12 stars**

BEST VIEWED: **Mar. / Apr.**

LOCATION: **Spring, center of chart**

SIZE ON THE SKY: ✋

ALPHA STAR: **Regulus**

DEEP SKY OBJECT: **R Leonis, variable star**

LEO IS A LARGE and easily recognized constellation visible from the Northern Hemisphere for the first half of the year. The lion sits in a rather empty portion of sky just beyond Ursa Major and between its fellow zodiacal constellations Cancer and Virgo. A sickle-shaped asterism represents the lifted head of the seated lion. Regulus, Leo's brightest star, is just south of the sickle and represents the lion's front legs. This is one of the easier constellations to construct mentally, as it resembles the classic image of the Sphinx. Spiral galaxies M65 and M66 near Theta Leonis are visible just beyond the base of the lion but require at least binoculars to detect.

History

The Sphinx, an icon of ancient Egyptian civilization, may have been modeled after this celestial beast.

INTERESTING OBJECTS in Leo

Named after the constellation from which they seem to originate, the LEONIDS are a meteor shower associated with comet Tempel-Tuttle. The meteors can be seen streaking across the sky when the Earth travels through the particles left by the passing comet and are best seen every year around November 17.

M66 (upper left) and M65 (lower left)

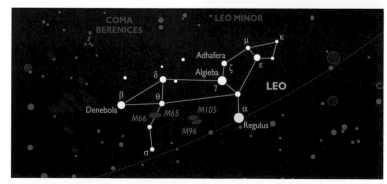

Leo Minor: The Small Lion

MAKEUP: 3 Stars

BEST VIEWED: Mar. / Apr.

LOCATION: Spring, center of chart

SIZE ON THE SKY: 🐾

ALPHA STAR: Alpha Leonis Minoris

DEEP SKY OBJECT: R Leonis Minoris

THIS GROUP did not belong to Ptolemy's original list of constellations. It is one of the more recent additions designated by Johannes Hevelius in the 17th century. It is a small and unobtrusive grouping, requiring a dark night to even be seen, as no star in the constellation is brighter than magnitude 3.8. In mid-northern latitudes, however, it is almost directly overhead during March and April, perched between Leo and Ursa Major.

Stars & Objects

The most notable of the Small Lion's dim stars is R Leonis Minoris, east of the three stars that form the constellation. It is a Mira star that will oscillate over a year between magnitude 7.1 and 12.6.

There is little of interest here for the backyard astronomer. There are no Messier objects, and its brightest deep sky object is an 11th-magnitude galaxy. Much of Leo Minor's characteristics are difficult to see with backyard equipment.

Leo Minor does hold an example of how astronomy makes advances through amateur astronomers. The Galaxy Zoo project encourages amateur astronomers to comb through pictures to try to classify galaxies. In one recent case, a Dutch observer noticed a strange green blob near spiral galaxy IC 2497. Professional astronomers are now theorizing about this body, which has become known as Hanny's Voorwerp—named for Hanny van Arkel, the schoolteacher who noticed it; *voorwerp* is Dutch for "object." Scientists think that the object may be a galaxy acting as a reflection nebula, relaying light coming from a quasar in IC 2497.

History

Because it is such a newcomer, Leo Minor has no associated Greek mythology. The ancient Egyptians, however, did recognize this small group of stars as the hoof prints left by a herd of gazelles as they fled from the pursuit of Leo (the Lion).

Main Stars

STARS	MAGNITUDE	DISTANCE (LY)
Alpha Leonis Minoris	3.4	98
Beta Leonis Minoris	4.4	146

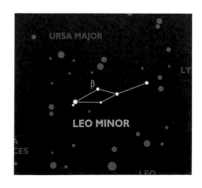

207

Lepus: The Hare

MAKEUP: 11 stars
BEST VIEWED: Jan. / Feb.
LOCATION: Winter, southeast quadrant

SIZE ON THE SKY: 🖐
ALPHA STAR: Arneb
DEEP SKY OBJECT: Hind's Crimson Star

THIS SMALL constellation has taken on multiple personalities throughout its history. It is recognizable less for its shape than for its position in the sky—just south of giant Orion and just west of the bright star Sirius in Canis Major. The Hare's head is just below Rigel, Orion's left foot. Midwinter sees Lepus highest in the sky, but it will still be close to the horizon for northern viewers, and dip below the southern horizon for part of the year.

Hind's Crimson Star, discovered by British astronomer J. Russell Hind, is a deep red variable that over roughly 14 months ranges from a magnitude 5.5 to a magnitude 11.7.

Mythology

The exact mythology behind Lepus is vague. It is commonly agreed that the hare is being hunted by Orion and his hounds Canis Major and Canis Minor.

Globular cluster M79

INTERESTING OBJECTS
in Lepus

Most globular clusters are found near the center of the Milky Way. GLOBULAR CLUSTER M79 however is found inside Lepus, a southern constellation on the opposite side of the sky. It has been suggested that this cluster was formed in the Canis Major dwarf galaxy.

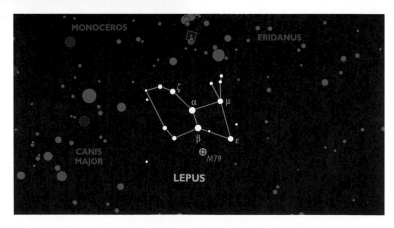

Libra: The Scales

MAKEUP: **8 stars**

BEST VIEWED: **June / July**

LOCATION: **Summer, southwest quadrant**

SIZE ON THE SKY: ✊

ALPHA STAR: **Zuben Algenubi**

DEEP SKY OBJECT: **Delta Librae**

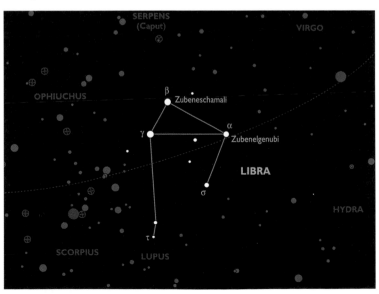

LIBRA'S three brightest stars form a triangle at the top of the scales and the two pans of the balance fall westward toward the bright star Antares in the constellation Scorpius. In the northern summer sky it will be just above the Milky Way, to the south, between its neighbors on the zodiac, Virgo and Scorpius.

The stars representing the pans of the scale were considered by Arab astronomers as part of the northern and southern claws of Scorpius—split off when Libra was created, apparently during the time of ancient Rome.

Of viewing interest is Delta Librae, found just west of the northern end of the beam. This star is an eclipsing variable whose 2.3-day cycle ranges from magnitude 4.9 to 5.9, all perfectly visible to the naked eye on a clear night.

Mythology

Libra is one of the 12 constellations of the zodiac and the only one not depicting a human or animal. According to myth, the scales are those held by Themis, the Greek goddess of justice.

SKYFACT

Zubenelgenubi and Zubeneschamali, Arabic names for two Libra stars, translate as the scorpion's left and right claws.

209

Lynx: The Lynx

MAKEUP: **7 stars**

BEST VIEWED: **Feb. / Mar.**

LOCATION: **Winter, northeast quadrant**

SIZE ON THE SKY: 🖐

ALPHA STAR: **Al Fahd**

DEEP SKY OBJECT: NGC 2419,Intergalactic Wanderer

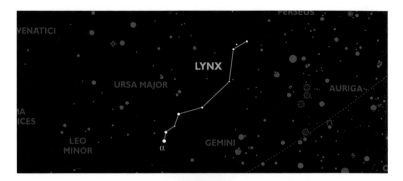

NOT MANY constellations are as inconspicuous as Lynx. Good luck if you are out to spot this constellation. It is said that Lynx earned its name because it requires the keen eyesight of a mountain cat to locate it. February and March are the best months to look for this constellation between Ursa Major and Gemini.

Stars & Objects

Lynx's lone deep sky object is the intriguingly named Intergalactic Wanderer (NGC 2419). It is one of the few named globular clusters, dubbed as it is by astronomer Harlow Shapley, who conjectured that it was so far away from the core of the Milky Way that it might break from its current loose orbit around the galaxy. A telescope of at least 10 inches of aperture is needed to spot it under good viewing conditions. This object is technically within the boundary of Lynx but is actually closest to the star Castor in Gemini than to any of the seven stars in Lynx.

History

Because this constellation is one of the groups of stars identified and named by Johannes Hevelius in the 17th century, it therefore lacks the ancient mythology associated with many of the other constellations. It represents a small North American wildcat rearing on its hind legs.

SKYFACT

Several of the constellations developed by Hevelius were shown publicly for the first time in his *Prodromus Astronomiae*, published after his death by his wife Elizabeth, a companion star watcher who is considered the first female astronomer.

Lyra: The Lyre

MAKEUP: 5 stars
BEST VIEWED: July / Aug.
LOCATION: Summer, center of chart

SIZE ON THE SKY: 〔♎〕
ALPHA STAR: Vega
DEEP SKY OBJECT: M57, Ring Nebula

LYRA is one of the delights of the northern summer sky, easy to envision and easy to spot. Its alpha star, Vega, is among the brightest stars in the sky, and is directly overhead from the mid-northern latitudes in the summer—a handy reference point for many other constellations and objects. With Deneb and Altair, Vega forms the Summer Triangle asterism. Located along one of Vega's borders, the Ring Nebula requires a 3-inch-aperture telescope and a high powered eyepiece to show its shape.

INTERESTING OBJECTS
in Lyra

RING NEBULA (M57) is located about 4,000 light-years away and is visible with a telescope. The famous ring is formed by gas that was emitted by a dying Sun-like star at the end of its existence. From our perspective on Earth, the nebula looks circular, but scientists have concluded that it is actually cylindrical in shape.

Mythology

A chilling love story accompanies this constellation. Apollo gave his son Orpheus the lyre and taught him to play captivating music. Despite being the object of many women's envy, Orpheus loved his wife Eurydice. She died before him and was sent to the underworld, but Orpheus was unyielding in his desire to bring her back to life.

The gods relented and decided to allow it, but Orpheus failed to follow their one admonition not to turn around to look back at his wife as they exited Hades. Having lost his love again, Orpheus refused all other advances and was killed by a discouraged group of young women. He was then reunited with his wife and as a tribute to their love, Zeus sent the Lyre to the sky.

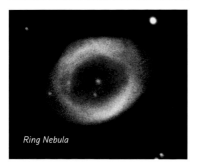

Ring Nebula

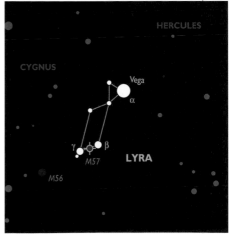

Monoceros: The Unicorn

MAKEUP: 8 stars
BEST VIEWED: Jan. / Feb.
LOCATION: Winter, southeast quadrant

SIZE ON THE SKY: 🖐
ALPHA STAR: Ctesias
DEEP SKY OBJECT: NGC 2244, Rosette Nebula

THIS IS ANOTHER of the more modern constellations created to fill gaps in the sky, suggested in 1624 by the German astronomer Jakob Bartsch. Though a dim group of stars, Monoceros fills a prominent gap in the sky, bounded by the very bright stars Procyon in Canis Minor, Betelgeuse in Orion, and Sirius in Canis Major. It is best viewed in the midwinter months of January and February when it is at its optimal position in the sky.

Stars & Objects

Though faint, with no stars above magnitude 3.5, the Unicorn's position on the Milky Way and amid these bright nearby stars provides many good reference points to help locate it in the night sky. The two stars closest to Betelgeuse represent the long head of the Unicorn. Its legs straddle the Milky Way. Finding its brighter neighbors may make spotting Monoceros an easier assignment for the first-time astronomer.

None of the deep sky objects in Monoceros are visible to the naked eye but there are a few of general interest which are visible with other equipment. The Rosette Nebula is a large emission nebula well viewed through a 10-inch-aperture telescope and a favorite target of many astrophotographers. Another highlight in the Unicorn is M50, an open cluster located between Sirius and Procyon, amid the Unicorn's outstretched legs. This open cluster is visible through binoculars.

History

Since it is so recent a creation, this constellation has little ancient mythology associated with it. There is mention in the Old Testament of the Bible to a "large wild animal" that decided to play in the rain rather than join Noah and the other animals on the ark to escape the great flood. Perhaps this story is why Bartsch decided to dedicate a constellation to the unicorn.

SKYFACT
Monoceros lies within the Winter Triangle asterism formed by neighboring stars Betelgeuse, Procyon, and Sirius.

Ophiuchus: *The Serpent Bearer*

MAKEUP: 14 stars
BEST VIEWED: June / July
LOCATION: Summer, center of chart

SIZE ON THE SKY: 🖐
ALPHA STAR: Rasalhague
DEEP SKY OBJECT: RS Ophiuchi

Globular cluster M14

THIS SOUTHERN constellation splits neighboring Serpens in two. The mythologies of the two constellations are as intertwined as their stars. Ophiuchus can be easily identified, lying just west of the Milky Way, above Scorpius and below Hercules. Rasalhague, its alpha star, represents the figure's head.

A rich constellation, Ophiuchus contains many deep sky objects. There is a series of globular clusters spread out in this constellation—M9, M10, M12, M14, M19, and M62—all visible through binoculars. RS Ophiuchi is a cataclysmic variable that ranges in magnitude from 11.8 to 4.3.

Mythology

Ophiuchus is thought to honor Asclepius, the god of medicine. The nearby snake, Serpens, taught the healer about the properties of plants, and he soon became skillful enough to raise the dead. His herbal remedies raised Orion after a fatal wound from Scorpius. Hades, god of the dead, became concerned and convinced Zeus to kill Asclepius. Now the god and his serpent watch from the sky.

INTERESTING OBJECTS
in Ophiuchus

BARNARD'S STAR, one of the nearest stars to Earth, is a 9.5-magnitude "star on the run." Named for the American astronomer E. E. Barnard who discovered it, this star is moving across the sky at 103 miles a second (166 km/s), the greatest proper motion of any known star.

213

Orion: The Hunter

MAKEUP: 20 stars
BEST VIEWED: Jan. / Feb.
LOCATION: Winter, southeast quadrant

SIZE ON THE SKY: ✋
ALPHA STAR: Betelgeuse
DEEP SKY OBJECT: Orion Nebula

THIS MAY be the most recognizable constellation in the sky, visible worldwide from its position straddling the celestial equator, and identified by different cultures across thousands of years. Its prominence earned three references in the makes the structure of a man in the constellation easy to imagine. The hunter's bright belt is made of three stars all around 2nd magnitude. His upraised, club-wielding arm sweeps into the Milky Way while his other hand holds a lion's pelt.

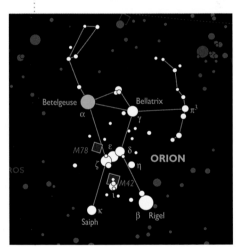

Betelgeuse
Bellatrix
π³
α
γ
M78
ε δ
ζ η
ORION
M42
ι
κ
Saiph
β Rigel

214

Bible, and it is mentioned in the *Iliad*, the *Odyssey*, and other classics of literature. To modern sky watchers, the hunter's position is ideal for use as a reference point: Three of the twenty-five brightest stars in the sky are found in this group, and their brightness

Bright Stars

Orion's main stars are among those that most easily stand out in the night sky. They can provide a way to orient yourself even amid the glare of city lights. Large, reddish tinted Betelgeuse is the alpha star and represents his right shoulder. It is a huge variable star that pulsates to a diameter between 300 and 400 times the size of our Sun. Blue-white Rigel, the hunter's left foot, is a magnitude 0.1 supergiant, 57,000 times as bright as the Sun.

The hunter's left shoulder is formed by another bright star, magnitude 1.6 Bellatrix (and if Orion's cultural centrality were in doubt, the name of this star would clarify it, appearing as an evil character in the Harry Potter series of books). Alnitak, Alnilam, and Mintaka—all around 2nd magnitude or less—form Orion's distinctive diagonal belt, one of the

SKYFACT

Orion is among the constellations to work its way into classic texts. It is mentioned three times in the Bible, for example, including this cosmic question from Job: "Can you bind the chains of the Pleiades, Or loose the cords of Orion?"

Backyard Guide to the Night Sky

Main Stars

STARS	VIEW	COLOR	MAGNITUDE	DISTANCE (LY)
Betelgeuse (alpha)	👁	Red	0.7	425
Rigel (beta)	👁	Blue	0.1	775
Bellatrix (gamma)	👁	Blue-white	1.6	240
Mintaka (delta)	👁	Blue	2.2	915
Alnilam (epsilon)	👁	Blue	1.7	915

characteristics by which it is most easily recognized.

Mythology

There are a few stories In Greco-Roman mythology associated with Orion. In one, Orion was a powerful hunter dealt a lethal sting by Scorpius—a reason for the two being placed, as they are, on opposite sides of the sky.

In another story, Artemis, the goddess of the hunt, admired Orion for his great skill as a hunter. Her twin brother, Apollo, grew jealous that his sister was showing favor to another man. He tricked her into shooting Orion with a deadly arrow. Upon discovering what she had done, Artemis immortalized Orion in the

Orion Nebula

stars. Now, the hunter also engages with the animals close to him, such as Taurus (the Bull) and Lepus (the Hare) along with his faithful hounds, Canis Major and Canis Minor.

INTERESTING OBJECTS in Orion

Orion includes an area of the Milky Way that features intense star production. Beneath the hunter's belt three stars form Orion's sword, and in the middle of them is the GREAT NEBULA (M42). Visible as a cloud patch to the naked eye, larger telescopes will show a gas cloud where star formation is taking place at a furious pace. At its center are the four stars known as the TRAPEZIUM. Another nebula, M43, lies just a bit to the north. For a challenge, try hunting south of Zeta Orionis for the HORSEHEAD NEBULA, a thick black cloud that takes on a horsehead shape, blotting out light from a nebular cloud behind it.

Pegasus: The Winged Horse

MAKEUP: **15 stars**
BEST VIEWED: **Sept. / Oct.**
LOCATION: **Autumn, center of chart**

SIZE ON THE SKY:
ALPHA STAR: **Markab**
DEEP SKY OBJECT: **M15, globular cluster**

CELESTIAL immortality for Pegasus is well earned. The winged horse has a large presence in Greek mythology. And despite its mild magnitude—alpha star Markab registers about 2.5—the constellation occupies a large position in the sky.

Stars & Objects

Pegasus is most easily seen in the fall, during the months of September and October. The constellation is found in the southern sky right by Andromeda, with whom Pegasus shares the star Alpharetz (Alpha Andromedae). This shared star joins with the three brightest stars of the horse to form the Great Square of Pegasus, an especially useful asterism in identifying other constellations in the regions close by.

The M15 cluster is one of the more prominent in the Northern Hemisphere, located on the far eastern border of the constellation. The dense grouping of stars is visible through binoculars, but a telescope is needed to see the main attraction—some 100,000 stars in a tight, sparkling ball.

Of general interest is Stephan's Quintet, a group of five galaxies visually close to one another near Pegasus's northern border. Four are in the process of colliding, while the fifth—based on calculations of its redshift—may be a foreground object closer to the Milky Way than the others.

Mythology

Pegasus is a well-known character in mythological literature. According

INTERESTING OBJECTS in Pegasus

Pegasus has been a repeated source of excitement in the search for EXOPLANETS—extrasolar planets, circling stars other than the Sun, a key target in the hunt for extraterrestrial life. In 1995, scientists detected an exoplanet around the star 51 Pegasi—the first exoplanet found around a stable star, as opposed to near a pulsar. Its presence was inferred by the effect of its gravity on the star. Two other exoplanets have since been discovered around other stars in the constellation.

But the most recent discovery is perhaps the most exciting. In November 2008, scientists released what are considered to be the first photographs of planets circling a star other than the Sun. In this case, what appear to be three planets were pictured circling the star HR 8799 in Pegasus—a veritable mini solar system. Like the other exoplanets in the constellation, the three news ones are all gas giants and not thought to be habitable. Still it was considered a breakthrough in the ongoing hunt for a planet outside our solar system that contains the right chemistry and composition to support life.

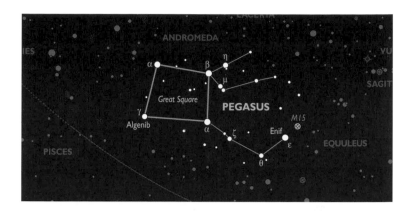

to myth he kicked open a mountain spring, awakening the waters of Hippocrene, which provided inspiration for generations of poets. Somewhat more familiar is the myth shared with the surrounding constellations Andromeda, Perseus, Cetus, Cepheus, and Cassiopeia. Refer to the complete story of Perseus and Andromeda, in which the winged horse served as Perseus's steed.

In another tale, Pegasus also served the Greek hero Bellerophon when he fought the Chimera. The hero was able to tame the flying horse with the help of the goddess Athena.

Stephan's Quintet

SKYFACT Pegasus is in a sense the offspring of Medusa. When the hero Perseus slew the monster, Pegasus was born of her blood.

Main Stars

STARS	VIEW	COLOR	MAGNITUDE	DISTANCE (LY)
Markab (alpha)		Blue-white	2.4	140
Scheat (beta)		Red	2.4	200
Algenib (gamma)		Blue	2.8	333
Enif (epsilon)		Orange	2.4	670

Perseus: The Hero

MAKEUP: **14 stars**
BEST VIEWED: **Nov. / Dec.**
LOCATION: **Autumn, northeast quadrant**

SIZE ON THE SKY: 🖐
ALPHA STAR: **Mirfak**
DEEP SKY OBJECT: **NGC 869 & 884, Double cluster**

VISIBLE for much of the year in the northern sky, the constellation Perseus straddles the Milky Way between Cassiopeia and Andromeda, characters from the same Greek myth in which he takes center stage as the hero.

Stars & Objects

The constellation itself is relatively easy to pick out for a beginning stargazer. With magnitude 1.8 Mirfak, the alpha star, at its center, and five other stars of magnitude 3 or greater, Perseus contains bright and notable signposts. Among them is Algol (Beta Persei), a well-known eclipsing variable that dims by a full magnitude for about ten hours every three days. Although visible for much of the year, the best times to view Perseus are in the late fall months.

Perseus contains a fascinating deep sky object. The Double cluster—NGC 869 and 884—gives the illusion of a large, connected field, but is in fact two separate, unassociated groups of stars. They are best studied with binoculars or a lower magnification eyepiece to provide a wide field of view of an area rich in double- and multiple-stars.

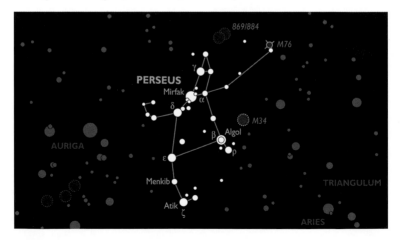

Backyard Guide to the Night Sky

SKYFACT

The variable star Algol represents the eye of Medusa in the story of Perseus. It also translates as "the ghoul" in Arabic and represented the eye of a vampire in Jewish tradition. Looks like the ancients suspected something was off-kilter about this star.

Perseids

The constellation is also the site of another notable night-sky event: a yearly meteor shower visible every summer beginning in late July. Perseus contains the radiant—or point of origin—for the annual Perseid meteor shower, which peaks on August 11 and 12 each year.

Meteor showers occur when the Earth passes through a trail of dust left by a traveling comet. The trail consists of debris and particles ejected by the comet. The Perseids come from the debris left by comet Swift-Tuttle, a long-period comet that last passed by Earth in 1992. The shower has been observed for almost 2,000 years, primarily by observers located in the Northern Hemisphere.

Mythology

Perseus, the half-mortal son of Zeus, is one of the more prominent Greco-Roman heroes. A favorite of Athena, he received many gifts from her, including a polished shield. Going into battle with the Gorgon Medusa, whose look would turn any human or animal into stone, Perseus used the shield as a mirror to avoid her direct glance. When he lopped off her head, the winged horse Pegasus was born of her blood and Perseus took the horse for his own.

In his second great feat, Perseus used Medusa's head as a weapon to help him rescue Andromeda from the sea monster Cetus. Even severed from her body, the Gorgon's stare turned Cetus into stone.

Main Stars

STARS	VIEW	COLOR	MAGNITUDE	DISTANCE (LY)
Mirfak (alpha)	👁	Yellow-white	1.8	590
Algol (beta)	👁	Blue-white	2.1	93
Alphecher (gamma)	👁	Yellow	2.9	225
Basel (delta)	👁	Red	3.0	530
Adid Australis (epsilon)	👁	Blue	2.9	538

Pisces: The Fish

MAKEUP: **17 stars**

BEST VIEWED: **Oct. / Nov.**

LOCATION: **Autumn, southeast quadrant**

SIZE ON THE SKY: 🖐🖐

ALPHA STAR: **Alrisha**

DEEP SKY OBJECT: **M74, spiral galaxy**

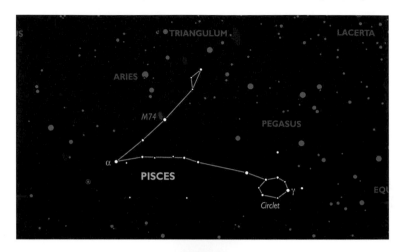

THIS ANCIENT constellation of the zodiac is in the shape of two fish attached by the tail to a long cord. Pisces is located near the square of Pegasus, overhead and to the southeast in October and November. The lengths of cord that hold the two fish together meet at alpha star Alrisha, just outside the constellation Cetus, forming a large V in the sky.

Five faint stars in the south of the constellation form the asterism known as the Circlet, representing the larger of the two fish. Just outside the cord attached to the smaller fish is M74, a faint spiral galaxy whose arms can be resolved with an 8-inch-aperture telescope. Van Maanen's star is a faint white dwarf to look for on the opposite side of the constellation—one of the few dead stars that can be seen in an 8-inch scope.

Main Stars

STARS	MAGNITUDE	DISTANCE (LY)
Alrisha (alpha)	3.9	139
Fum Al Samakah (beta)	4.5	495
Simmah (gamma)	3.7	130
Kullat Nunu (eta)	3.7	294

Mythology

The ancient Greeks believed that the goddess Aphrodite and her son Eros changed themselves into fish to escape the sea monster Typhon. The cord holds mother and son together as they swim.

Piscis Austrinus: The Southern Fish

MAKEUP: 11 stars
BEST VIEWED: Sept. / Oct.
LOCATION: Autumn, southwest quadrant

SIZE ON THE SKY:
ALPHA STAR: Fomalhaut
DEEP SKY OBJECT: None

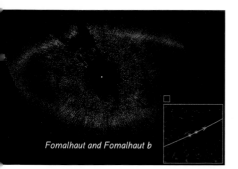

Fomalhaut and Fomalhaut b

PISCIS AUSTRINUS

THIS SOUTHERN constellation—low to the horizon even at its highest for northern mid-latitude observers—still has the advantage of being marked by Fomalhaut, a 1st-magnitude star lying in a relatively dim section of sky.

Piscis Austrinus is easily identified in the fall just south of Aquarius and east of Capricornus—its companions in this "watery" section of the sky. Fomalhaut is the most distinct object in this constellation, the 18th brightest star to the naked eye, and just 25 light-years from Earth. Fomalhaut is a relatively young star, between 100 and 300 million years old. Its life span is estimated to be as long as a billion years. It is twice as large as the Sun and burns between 14 and 17 times as bright.

Mythology

Piscis Austrinus is one of the original 48 constellations identified by Ptolemy. Early artwork of this constellation depicts the sea creature with an open mouth, drinking water poured from the jug of Aquarius. The ancient Egyptians saw this group of stars as a fish, indeed the name Fomalhaut is derived from the Arabic translation for "mouth of the fish." One Assyrian story suggests that the fish represents the creature Oannes, who was a teacher during the day but transformed at night. In Greek mythology, Piscis Austrinus is said to be the sire of the two fish in the constellation Pisces.

INTERESTING OBJECTS in Piscis Austrinus

A breakthrough observation of the young star FOMALHAUT was made in November 2008 when the Hubble telescope photographed a planet orbiting the star. Scientists named the planet Fomalhaut b; it was the first confirmed extrasolar planet detected via direct imaging.

Puppis: The Stern

MAKEUP: 12 stars

BEST VIEWED: Feb. / Mar.

LOCATION: Winter, southeast quadrant

SIZE ON THE SKY: 🤚

ALPHA STAR: Suhail Haddar

DEEP SKY OBJECT: M46, open cluster

THIS SOUTHERN constellation is another floating in the Heavenly Waters, an area of sky that contains Eridanus, Pisces, Aquarius, and the three parts of the disassembled ship *Argo*. The *Argo* was the ship that carried Jason and his 50 Argonauts on their journey to steal the Golden Fleece that belonged to Aries (the Ram). Its sail is now seen in Vela, its keel in Carina and its anachronistic compass in Pyxis.

Stars & Objects

Formerly one large constellation, Jason's mythical ship is the only one of Ptolemy's original constellations to be so thoroughly revised and divided. Puppis represents the stern of the mighty craft. Today, Puppis does not have a true alpha star because the stars were not relabeled when the large constellation Argo was split into pieces. Thus Carina kept the original alpha and beta stars from Argo, while the brightest in Puppis was the zeta star of the original constellation.

From mid-northern latitudes, Puppis will be close to the southern horizon. To locate it, find the bright stars Sirius and Adhara of Canis Major. Trace a line south directly to Suhail Hadar, the brightest star in Puppis and a notable, bright red variable.

Among the deep sky objects in Puppis is the open cluster M46. It appears as a faint cloud about the size of the Moon through binoculars but resolves into a sparkling field of stars through a telescope. This small constellation also include two stars worth singling out: Second-magnitude Zeta Puppis is one of the Milky Way's largest stars, a blue supergiant. L2 Puppis is among the most luminous of the red variables, changing from magnitude 2.6 to 6.2 over about five months.

INTERESTING OBJECTS in Puppis

The GUM NEBULA is the largest known nebula seen from Earth and the remnant left by a supernova that exploded about a million years ago. It spans 40 degrees of the sky, across the constellations Puppis and Vela, two far-southern constellations that are largely out of sight for northern viewers. The nebula's near side lies about 450 light-years away.

Sagitta: The Arrow

MAKEUP: 6 stars
BEST VIEWED: Aug. / Sept.
LOCATION: Summer, southeast quadrant

SIZE ON THE SKY: 👍
ALPHA STAR: Sham
DEEP SKY OBJECT: M71, globular cluster

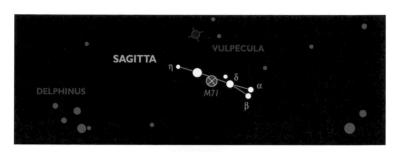

THIS SMALL but distinct constellation is inside the Milky Way on a direct line traced between Altair of Aquila and Deneb of Cygnus. For centuries this has been recognized as an arrow—obvious from its shape. Sagitta is one of the smallest constellations and one of Ptolemy's original 48 constellations. Located very close to the celestial equator, Sagitta can be seen from almost any location on Earth.

The entirety of this small constellation lies within the Summer Triangle asterism of Altair, Deneb, and Vega. This locates the constellation high in the sky during the summer months, when it is best seen it the mid-northern latitudes. A deep sky object of interest is M71, a globular cluster easily seen through binoculars in the center of the arrow's shaft. The bright cluster is located about 12,000 light-years from Earth and measures about 27 light-years across.

Mythology

Although the constellation is small and not very bright, many cultures—including the Greeks, Romans, Persians, and Hebrews—designated this group of stars as an arrow.

In Greco-Roman myth, Sagitta is said to have been several different arrows, including the one used by Cupid in the art of conjuring love, Apollo in slaying the Cyclops, and Hercules in dispatching the Stymphalian Birds. It may also have been an arrow from the bow of the archer Sagittarius in the hunt for one of the many nearby animals. This assertion is controversial as Chiron the centaur, the figure associated with Sagittarius, was known for his peaceful demeanor.

Main Stars

STARS	MAGNITUDE	DISTANCE (LY)
Sham (alpha)	4.4	473
Beta Sagittae	4.4	466
Gamma Sagittae	3.5	274

223

Sagittarius: The Archer

MAKEUP: 22 stars

BEST VIEWED: July / Aug.

LOCATION: Summer, southeast quadrant

SIZE ON THE SKY: 🖐

ALPHA STAR: Rukbat

DEEP SKY OBJECT: M8, Lagoon Nebula

THE ARCHER Sagittarius is a centaur—half man, half horse—located at the widest band of the Milky Way and offering a window to the center of the galaxy. Bordered by zodiac neighbors Scorpius and Capriconus, Sagittarius appears in mid to late summer in the southern sky as he pursues his supposed prey, Scorpius, through the galaxy. To find Sagittarius locate Vega and look for the group of eight bright stars that form the core of this constellation and mark the centaur's broad shoulders.

Another way to spot Sagittarius is to look for two well-known asterisms that fall within the constellation. The eight central stars form the shape of a teapot, with the easternmost four forming the handle, and a triangular set of three to the west forming the spout. By itself, the western rectangle is known as the Milk Dipper. You can think of the Milky Way as steam coming from the Teapot's spout.

Stars & Objects

Among the more well-known star forms, Sagittarius is a prominent example where the Bayer system for naming stars does not reflect the order of magnitude. Sometimes the cataloging is off, either because of mistakes or later changes in constellation borders, and the brighter star turns out to be the beta or gamma star. Sagittarius is a prime example. Sagittarius has plenty of candlepower—eight of its stars are brighter than magnitude 3. But the alpha star, Rukbat, is only the 14th brightest, at magnitude 4. The brightest star is Kaus Australis—Epsilon Sagitarii—at magnitude 1.8. The second brightest star, meanwhile, is Sigma Sagitarii—also known as Nunki, prominent enough to be included by the Babylonians on their Tablet of the 30 Stars.

INTERESTING OBJECTS in Sagittarius

The broad portion of the Milky Way where Sagittarius rests is oriented toward the core of the galaxy and many globular star clusters and nebulae. It is a part of the sky where binoculars or a telescope will let you skip among a half dozen or more closely situated deep sky objects. The most striking of these is the GREAT SAGITTARIUS STAR CLUSTER, located northwest of the lid of the Teapot; it is visible with binoculars, but an 8-inch or larger telescope will show an overwhelming number of individual stars. On the other side of the lid, the LAGOON NEBULA (M8), is visible to the naked eye on dark nights. Further north the OMEGA NEBULA (M17), likened to a horseshoe or swan, is within reach of binoculars. The TRIFID NEBULA (M20) is a stellar nursery, full of embryonic stars.

Main Stars

STARS	VIEW	COLOR	MAGNITUDE	DISTANCE (LY)
Rukbat (alpha)	👁	Blue-white	4.0	170
Arkab Prior (beta-1)	👁	Blue-white	4.0	139
Arkab Posterior (beta-2)	👁	White	4.3	139
Alnasl (gamma)	👁	Orange	3.0	96
Kaus Media (delta)	👁	White	2.7	305
Kaus Australis (epsilon)	👁	Blue	1.8	145

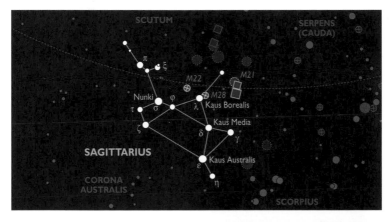

Mythology

The ancient Greeks identified Sagittarius with the centaur Chiron, who also appears in the constellation Centaurus. Chiron, however, was a wise and peaceful creature—a tutor of Achilles, Hercules, and others—in contrast to this archer's drawn bow. Others claim that Sagittarius is a satyr, a two-legged descendant of the Greek god Pan.

Trifid Nebula

SKYFACT

Tea is popular in Arab cultures today, but the ancient Arab astronomers did not associate Sagittarius's stars with the obvious appliance. To them the bright group of stars that forms the teapot represented ostriches on their way to drink from the Milky Way.

Scorpius: The Scorpion

MAKEUP: **17 stars**

BEST VIEWED: **July / Aug.**

LOCATION: **Summer, southwest quadrant**

SIZE ON THE SKY: 🖐

ALPHA STAR: **Antares**

DEEP SKY OBJECT: **M6, Butterfly cluster**

SCORPIUS is one of the most evocative images in the night sky. Two of the 25 brightest stars in the sky are in this constellation and they are surrounded by countless other brilliant stars and deep sky objects. For northern observers, it will be spotted best in the summertime toward the south along the Milky Way, between zodiac partners Sagittarius and Libra.

Unlike many constellations, Scorpius very much resembles its namesake animal—from its wide, pincer-armed head to a torso represented by bright Antares and a twisted tail that ends with the bright star Shaula.

Brightest Stars

For novice star watchers, Scorpius is among the easier constellations to

INTERESTING OBJECTS in Scorpius

Scorpius's section of the sky is teeming with star clusters. GLOBULAR CLUSTER M4 lies just west of Antares. The BUTTERFLY CLUSTER (M6) is a large open cluster that, with the help of binoculars or a telescope, unfolds into the shape of a butterfly. M7 is a large open cluster whose individual stars will come into view with binoculars. Over 100 stars occupy this cluster. Ptolemy recognized this group of stars and it is thus named PTOLEMY'S CLUSTER. It is visible just east of Shaula. SCORPIUS X-1 is faint at 13th magnitude—it will take some skill to find. A binary star, only one member of the pair can be seen: The partner is a white dwarf, neutron star, or black hole drawing gas away for its neighbor.

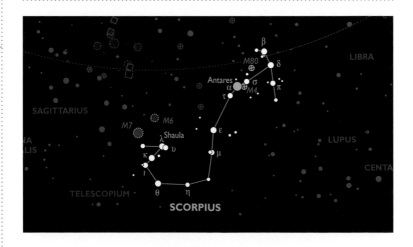

Backyard Guide to the Night Sky

Main Stars

STARS	VIEW	COLOR	MAGNITUDE	DISTANCE (LY)
Antares (alpha)	👁	Red	1.0	600
Acrab (beta)	👁	Blue-white	2.6, 4.9	530
Zubenhakrabi (gamma)	👁	Red	3.3	290
Dschubba (delta)	👁	Blue-white	2.2	400
Wei (epsilon)	👁	Red	2.3	65

recognize, both because of its shape and because of the 1st-magnitude star, Antares, that sits at its heart. Antares was known to the Romans as Cor Scorpionis, or the "heart of the scorpion."

A red supergiant, Antares is about 300 times as large as our Sun, and is nearing the end of its stellar life. It has been one of the stars more closely watched by humans: To the Chinese, the star is Ming T'ang or the "emperor's council hall." Persians considered Antares one of the guardians of the heavens and called it Satevis.

Mythology

This celestial shape has been significant to numerous civilizations. Chinese astronomers saw a mighty dragon. Ancient Greek mythology recognizes Scorpius as the animal that bested the hunter Orion; he was able to inflict the fatal wound on the hero's leg, which

is now visible in the sky beginning at Rigel. The hunter and scorpion are on opposite sides of the heavens to keep them separated—with Scorpius rising in the east in the spring just as Orion sets in the west after a winter in the northern sky.

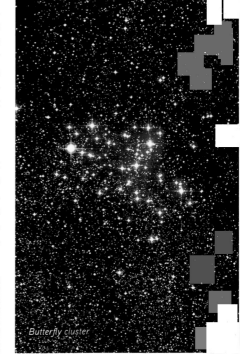

Butterfly cluster

SKYFACT

The name Antares is Greek for "rival of Mars"—testament to the red star's color and prominence in the sky.

Scutum: The Shield

MAKEUP: **4 stars**

BEST VIEWED: **July / Aug.**

LOCATION: **Summer, southeast quadrant**

SIZE ON THE SKY: (☝)

ALPHA STAR: **Alpha Scuti**

DEEP SKY OBJECT: **M11, Wild Duck cluster**

SCUTUM IS another one of the constellations that was named by Johannes Hevelius in the late 17th century. It is a modern constellation, but its name and shape invoke historical events—perhaps the only constellation with contemporary (at least to Hevelius) political overtones.

Stars & Objects

The shield can be found on the Milky Way between Sagittarius, Aquila, and the lower half of Serpens. Scutum is not large nor does it have any very bright stars, but its distinguished company makes it a good viewing target in the Northern Hemisphere in the summer months of July and August.

Though small, the constellation's position in the Milky Way makes it home to one of the galaxy's highlights. The Wild Duck cluster (M11), so named because it is reminiscent of a thick flock of waterfowl, is clearly visible through binoculars southwest of the constellation's northernmost star (Beta Scuti). Just an 8-inch telescope will start to show the cluster's thousands of glittering stars.

History

The original name for this constellation was Scutum Sobiescanum, which translates as "shield of Sobieski." John Sobieski was the king of Poland and commander of the forces that defeated the Ottoman Empire in the critical 1683 Battle of Vienna. King John was called upon to lead 70,000 men against an army twice that size. It might seem odd for Hevelius to have named a constellation after a man who was still alive. But Sobieski—along with repulsing what was considered a major threat to Europe—was also one of the astronomer's financial backers.

INTERESTING OBJECTS in Scutum

Between 1999 and 2004, the Chandra X-ray Observatory captured incredible images of the SUPERNOVA G21.5-0.9, which is located about 20,000 light-years away. Astronomers estimate that a star well over ten times more massive than the Sun produced this explosion. Though the supernova has been known for a number of years, Chandra was able to highlight the outer shell of ejected x-ray material.

SCUTUM

Sextans: The Sextant

MAKEUP: **3 stars**
BEST VIEWED: **Mar. / Apr.**
LOCATION: **Spring, southwest quadrant**

SIZE ON THE SKY: (♍)
ALPHA STAR: **Alpha Sextantis**
DEEP SKY OBJECT: **NGC 3115, Spindle galaxy**

S EXTANS URANIAE's full name has been short-ened over the years to just Sextans. This small constel-lation is a 17th-century creation of astronomer Johannes Hevelius and fills out the sky near Leo's portion of the ecliptic.

Stars & Objects

This is a relatively inconspicuous con-stellation. It is composed of a handful of stars and none greater than a mag-nitude 4.5. Sextans can be found in the south occupying the space between Hydra and Leo.

Observing Sextans for the first time may be challenging for the begin-ning astronomer. Most of its stars—even the brightest—are difficult to see

unaided. A clear Moonless night and a dark location away from city lights and any other light-pollution sources would yield the best viewing condi-tions for seeing the Sextant.

Sextans's small patch of sky does contain one notable deep sky object—the Spindle galaxy, viewed edge on as a flattened disk. Amateur telescopes will reveal a rewarding, similar view seen in photographs. The constellation is also the radiant point of the daytime Sextanid meteor shower, which recurs every four years (due next in September 2009).

History

Sextans was only recently granted individual attention in the sky. Until the late 1600s, this group of stars represented a fin on the back of the giant multiheaded monster Hydra. Johannes Hevelius named the con-stellation in 1687 for the astronomical instrument used to measure the posi-tions of the stars. The minor constel-lation served as a memorial of sorts to Hevelius's own sextant, which (along with his other astronomical instru-ments) was destroyed in a fire in Sep-tember 1679.

SKYFACT

Sextans's stars are not very bright and most are difficult to see with the naked eye, but that did not stop ancient Chinese astronomers from appointing one of its fainter stars, the one they called Tien Seang, to represent the Minister of State in Heaven.

Serpens: The Serpent

MAKEUP: 8 stars (Caput) & 6 stars (Cauda)
BEST VIEWED: June / July
LOCATION: Summer, southern half of chart

SIZE ON THE SKY:
ALPHA STAR: Unukalhai
DEEP SKY OBJECT: M16, Eagle Nebula

THIS CONSTELLATION winds over the shoulders of Ophiuchus, the Serpent Bearer, toward the south. These two groups of stars were once one constellation, along with Ophiuchus, but they have since been split. The snake is now the only divided constellation. Serpens Caput represents the head, and Serpens Cauda represents the tail. Serpens Cauda arcs between Altair and Antares. From there, the broad shoulders of Ophiuchus come into view, while the triangular head of Serpens Caput is just on the other side.

The Eagle Nebula (M16) in Serpens Cauda is a combination nebula and star cluster. The two can both be spotted in a telescope with at least an 8-inch aperture. Smaller equipment will show off the cluster.

Mythology

The snake in this constellation is supposed to have taught the medicine god Asclepius. The god used the same principles to restore the dead to life. This was unacceptable to the gods, who killed Asclepius and sent him and the serpent to the sky.

SKYFACT

The symbol of the modern striped medicine pole was adapted from the ancient Greek symbol of the rod of Asclepius. The god's staff is shown with a serpent winding its way around from top to bottom. On the modern pole, the red spiral was meant to represent the healer's serpent.

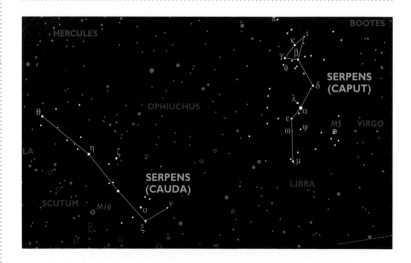

Main Stars

STARS	VIEW	COLOR	MAGNITUDE	DISTANCE (LY)
Unukalhai (alpha)		Orange	2.7	73
Chow (beta)		White	3.7	153
Tang (eta)		White	3.2	62
Leiolepis (mu)		White	3.5	156

Eagle Nebula

INTERESTING OBJECTS in Serpens

THE EAGLE NEBULA (M16) is an active region of star formation located in Serpens Cauda, about 7,000 light-years away. It was discovered independently by Philippe Loys de Chéseaux between 1745 and 1746 and then by Charles Messier in 1764. Messier described the stars in the nebula as "enmeshed in a faint glow." The nebula has dark pillars of dense material rising in its center; these can be seen with a 12-inch telescope. HOAG'S OBJECT lies about 600 million light-years away toward the constellation Serpens. Surrounded by a ring of hot, blue stars, Hoag's object is a ball of old, red stars; there is nothing between these two sets of stars. Astronomers are not sure how this object formed, though similar objects have been found. They have been given the name ring galaxies. The Hubble Space Telescope captured an image of destruction and star formation that will continue for billions of years. Known as the SEYFERT'S SEXTET, the Hubble image shows what looks like six galaxies interacting. In reality, two are passive observers while the other four collide. Gravity is acting strongly on these tightly packed galaxies, and is distorting their shapes.

Taurus: The Bull

MAKEUP: **13 stars**
BEST VIEWED: **Jan. / Feb.**
LOCATION: **Winter, southwest quadrant**

SIZE ON THE SKY: 🖐
ALPHA STAR: **Aldebaran**
DEEP SKY OBJECT: **M45, Pleiades; M1, Crab Nebula**

A Native American tale tells of seven young sisters who took a walk and lost their way. Now they stay in the sky as a reminder for children not to stray too far from home. Binoculars will help reveal the cluster's 500-plus stars.

Another interesting object in this constellation is the Crab Nebula, the first object named in the Messier list. In 1054, ancient astronomers witnessed the death of a star, and the nebular remnant of the supernova can still be seen just above the southern horn of the Bull.

TAURUS is a constellation of the zodiac at its northern peak in winter. It is easy to locate because Orion's belt points toward Aldebaran, the alpha star and the eye of the bull.

The Pleiades star cluster farther to the northeast marks the animal's shoulder. Also known as the Seven Sisters, the Pleiades is one of the more easily identified objects in the night sky. A line traced from Betelgeuse, in Orion, through Aldebaran will bring the dazzling group into view. Myths of many cultures have been spun around the seven bright stars in this cluster.

Mythology

Numerous myths accompany this constellation. The ancient Egyptians associate the Bull with Osiris, god of life and fertility. According to Greek mythology the Bull is one of Zeus's disguises, used to capture Europa and bring her across the ocean to the continent that now carries her name. To some, Taurus represents the golden calf formed by the idling followers of Moses while he was receiving the Ten Commandments.

INTERESTING OBJECTS in Taurus
In 1758, the CRAB NEBULA prompted Charles Messier to begin his extensive catalog of deep sky objects that resembled comets, earning it the title M1. This is not ill deserved: The supernova that produced the Crab Nebula appears everywhere from records of the ancient Chinese to cave art in the American Southwest. The supernova was said to have been visible to the naked eye for almost two years and visible during daylight for almost a month.

Triangulum: The Triangle

MAKEUP: 3 stars
BEST VIEWED: Nov. / Dec.
LOCATION: Autumn, southeast quadrant

SIZE ON THE SKY: (⌐)
ALPHA STAR: Mothallah
DEEP SKY OBJECT: M33, Pinwheel galaxy

ONCE the rather faint stars are identified just east of Andromeda, the small Triangle is easily identified. It is one of Ptolemy's original 48 constellations and has many different associations across many different cultures. Located between Andromeda, Perseus, and Aries, the constellation is best seen from the Northern Hemisphere in December, when it is high and central in the sky.

Stars & Objects

The most distinct object in Triangulum is the Pinwheel galaxy (M33), part of the local group of galaxies that includes the Milky Way. M33 is three million light-years away from Earth. Like our own galaxy, the Pinwheel is a spiral galaxy, and viewed from Earth's head-on perspective the shape is distinct. The Pinwheel is bright enough to be spotted as a slight glow on dark nights with very good viewing conditions.

But a telescope with a wide field of view is needed for the full pinwheel effect.

In 2007, a black hole about 15.7 times the mass of the Sun was discovered in the galaxy by the Chandra X-ray Observatory. Dubbed M33 X-7, the black hole orbits a companion star, which it eclipses every 3.5 days. It is the most massive black hole ever discovered. Scientists predict that the black hole's partner star will eventually go supernova and then there will be two black holes in the galaxy.

Mythology

Like many constellations, Triangulum has represented different things to various groups of astronomers—the ancient Hebrews appear to have applied the name that stuck, likening it to the small musical instrument. Before it took its contemporary name, ancient Greeks paid attention to this indistinct group of stars because it looked like their letter *delta* (Δ)—and for some time was known as Deltoton.

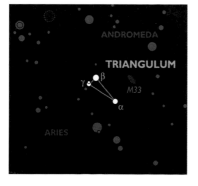

Main Stars

STARS	MAGNITUDE	DISTANCE (LY)
Mothallah (alpha)	1.9	415
Beta Trianguli	2.9	40
Gamma Trianguli	2.9	183

Ursa Major: *The Great Bear*

MAKEUP: **20 stars**

BEST VIEWED: **Mar. / Apr.**

LOCATION: **Spring, center of chart**

SIZE ON THE SKY:

ALPHA STAR: **Dubhe**

DEEP SKY OBJECT: **Mizar & Alcor**

URSA MAJOR, or the Great Bear, is one of the dominant shapes in the northern sky, including the seven-star asterism known as the Big Dipper (the Plow in Britain). It is one of the most ancient constellations, associated with stories that have cut across many cultures—and alternately turned the Great Bear into a chariot, a horse and wagon, a team of oxen, or (from the ever imaginative Egyptians) a hippopotamus.

See "Five Coolest Things in the Sky," pp. 46–47

Backyard Guide to the Night Sky

INTERESTING OBJECTS in Ursa Major

A tour of Ursa Major begins with the stars ALCOR and MIZAR, which lie close together in what has been determined to be an illusory pairing. Visible as two stars to a sharp naked-eye observer, the "nearness" of the two is a trick of alignment: They are not actually bound in a binary star system. With a small telescope, however, you'll discover that Mizar does have a companion, with Mizar B orbiting about 14 arc seconds away, the first true binary to be discovered.

The bear also hosts a pair of GALAXIES—M81 and M82—that can be spotted together in binoculars. M81 is a spiral galaxy that may give a sense of what the Milky Way would look like from the outside, while M82 may be undergoing a series of explosions caused by a collision with its neighbor. Also in this constellation is the OWL NEBULA (M97). A 3-inch-aperture telescope is needed to find this planetary nebula, but a much larger scope—12 inches or more—is needed to see the shape that provided the name.

Main Stars

STARS	VIEW	COLOR	MAGNITUDE	DISTANCE (LY)
Dubhe (alpha)	👁	Orange	1.8	124
Merak (beta)	👁	White	2.4	79
Phecda (gamma)	👁	White	2.4	84
Megrez (delta)	👁	White	3.3	81
Alioth (epsilon)	👁	White	1.8	81

Galaxy M81

Stars & Objects

Its size and distinct shape make a useful reference for locating other objects in the sky. From the north, over the course of the year, the Great Bear appears to run in a circle with its back to Polaris, the North Star. March and April mark the season when the Bear is highest in the sky. The easily identified Big Dipper represents the rear torso and tail of the Bear, with the other stars mapping out its long nose and legs: It's among the few constellations with a close-to-literal overall shape.

Mythology

In Greek mythology the Great Bear represents the nymph Callisto who was transformed into a bear by Hera, enraged after discovering that her husband Zeus had impregnated her. Callisto's son, Arcas, mistakenly tried to kill her while hunting. Zeus intervened and placed them in the sky.

Some Native American tribes believe the cup of the dipper represents a bear and the stars in the handle represent warriors that pursue the animal. In autumn, when the leaves turn, the constellation is low in the sky and the trees are thought to be stained red with the blood of the injured bear.

235

Ursa Minor: The Small Bear

MAKEUP: **7 stars**

BEST VIEWED: **Year-round**

LOCATION: **All star charts**

SIZE ON THE SKY: **🖐**

ALPHA STAR: **Polaris**

DEEP SKY OBJECT: **None**

THIS NORTHERNMOST constellation is visible year-round to star watchers in the Northern Hemisphere. Its value as a navigational tool was recognized long ago, and travelers still look to the Little Dipper to locate Polaris, the North Star. Other stars in the sky rotate about the celestial pole, making Polaris and the Little Dipper reliable constants in the night sky not only for travelers, but for those navigating the stars. Find Ursa Minor and Polaris, and you have a sense not just of direction, but of position—the North Star's distance above the horizon is roughly equivalent to latitude. Earlier cultures even used it as a clock, marking time as the constellation swung around the seemingly fixed point of the North Star.

Polaris is the most notable object in this small constellation. Through the recorded history of astronomy, the celestial pole has shifted. It is still approaching Polaris and will reach its closest around the year 2100. The pole will continue to move past it and through a succession of new polestars, first in Cepheus, then Cygnus, and eventually Vega about 12,000 years from now. Check the predawn sky in late December for the Ursid meteor shower erupting from Ursa Minor at about ten meteors per hour at its peak.

Mythology

Ursa Minor is the mythic child of Callisto, or the Great Bear, placed into the sky by Zeus. According to Native American legend, people were frightened that they would not be able to find their way on Earth and so asked the gods for help. They were granted the polestar to guide them.

URSA MINOR

Polaris
α
ζ
η β
Kochab
γ

Main Stars

STARS	MAGNITUDE	DISTANCE (LY)
Polaris (alpha)	2.0	430
Kochab (beta)	2.1	126
Pherkad (gamma)	3.1	480
Yildun (delta)	4.4	183

See "Getting Oriented," pp.14-15

Vulpecula: The Fox

MAKEUP: **3 stars**
BEST VIEWED: **Aug. / Sept.**
LOCATION: **Summer, southeast quadrant**

SIZE ON THE SKY: 🦊
ALPHA STAR: **Alpha Vulpeculae**
DEEP SKY OBJECT: **M27, Dumbbell Nebula**

THE FOX is found high in the late summer sky, roughly in between the bright stars Altair in Aquila and Vega in Lyra, inside the band of the Milky Way. It is not an easy constellation to pick out because it is far dimmer than its surroundings.

Stars & Objects

Vulpecula's stars may be faint, but within the Fox is a fine example of a planetary nebula, M27, the Dumbbell Nebula. Discovered in 1764 by Charles

Messier, M27 was the first reported planetary nebula. A good telescope will reveal the shape behind the object's nickname, but it can be spotted as a faint, fuzzy patch through binoculars just north of the brightest star in Sagitta, the constellation below.

Another elegant, even startling, find in Vulpecula is the Coathanger, an asterism on the border with Sagitta. On a dark night it is visible to

the naked eye, but binoculars resolve it into individual stars—a straight line of six with a "hook" of four in the middle. First-time viewers are invariably jolted to see such an artificial-looking arrangement of stars.

History

What this constellation lacks in mythology or character, it makes up for in unique astronomical interest. Johannes Hevelius invented the constellation and named it Vulpecula cum Anser, which means "the fox with the goose." The name has since lost the second creature, but older depictions of the constellation show a fox running as it clasps a struggling goose in its jaws.

INTERESTING OBJECTS
in Vulpecula

In 1764, Charles Messier added nebula M27 to his list of objects that are not comets, but it was John Herschel who nicknamed it the DUMBBELL NEBULA. M27 is what is known as a gaseous emission nebula. This is the product of a Sun-like star running out of fuel to keep it burning. The nebula is the outer layer of the star as it expands into space. In addition to the first planetary nebula, Vulpecula also contains the first observed radio pulsar, or rotating star— PSR B1919+21.

Virgo: The Virgin

MAKEUP: **13 stars**

BEST VIEWED: **May / June**

LOCATION: **Spring, southeast quadrant**

SIZE ON THE SKY:

ALPHA STAR: **Spica**

DEEP SKY OBJECT: **M104, Sombrero galaxy**

VIRGO is another of the zodiacal constellations—the only one representing a woman. Spica, located in the ear of wheat she

Stars & Objects

Like the bounty of wheat she holds in her hand, Virgo is among the richest areas of the sky to explore—prime territory to

holds in her left hand, is the brightest star in the constellation and is helpful in visualizing Virgo's full shape. This is a 1st-magnitude star and by far the brightest one in that part of the sky— and thus easy to find. It is a key star when star hopping: From the end of the handle of the Big Dipper, move south in the "arc to Arcturus," in the constellation Boötes, then "speed on to Spica" directly beneath it.

spot galaxies, and the location of what is the most distant night-sky object that backyard observers are likely to spot by themselves. The quasar 3C273 Virginis is some three billion light-years away. A minimum 8-inch telescope is needed to spot it.

Just east of Spica is the distinctive Sombrero galaxy (M104), marked by the dark band across its middle that acts as the brim of the hat which provided the galaxy's name.

To the north, near the edge of the constellation's border, is the heart of what is known as the Virgo-Coma cluster— some 13,000 star fields scattered among Virgo and nearby Coma Berenices.

SKYFACT

"Arc to Arcturus, speed on to Spica"—amateur sky watchers use this mantra to help find these bright landmark stars.

On May 11, 1781, Chares Messier added the SOMBRERO GALAXY (M104) to his catalog. The bright and spectacular galaxy is one of the largest in Virgo. The large bulge in the center gives the impression of a sombrero floating in space.

Spanning more than five degrees of the sky, an area about ten times that of the full moon, is the VIRGO CLUSTER of galaxies. More than 2,000 galaxies are grouped together and are the closest such group to our own Milky Way. The gravitational force of this cluster is so strong that it is slowly pulling our galaxy toward it. Through studying the movement of the galaxies, scientists realized that this region contains more dark material than visible material.

NGC 5426 and NGC 271 (known collectively as Arp 271) are galaxies of very similar sizes that are passing extremely close to each other. Normally, if two galaxies come into such close proximity, the smaller one will be consumed by the larger. Because these are both spiral galaxies with compact cores of about the same size, they are likely to survive the encounter; however, new stars will form in the wake of gas.

Though an 8-inch aperture telescope will be needed to capture the most detailed views, this is an area that will reward casual trolling with even a smaller piece of equipment. Some of the more prominent sights include elliptical galaxy M49; the twin sight of M84 and M86, ellipticals that will be spotted in the same field of a lower-powered eyepiece; and M87, an elliptical that may never render more than a fuzzy view, but professional equipment has shown it to host thousands of globular clusters.

Virgo cluster

Main Stars

STARS	VIEW	COLOR	MAGNITUDE	DISTANCE (LY)
Spica (alpha)	👁	Blue	0.1	260
Zavijava (beta)	👁	White	3.6	36
Porrima (gamma)	👁	White	2.7	38
Auva (delta)	👁	Red	3.4	202
Vindemiatrix (epsilon)	👁	White	2.9	102

MOST OF the world's population lives in the Northern Hemisphere, but some of the best celestial sights are visible only in the south. From 50 degrees south down to the South Pole, you'll be treated to glorious clusters, galaxies, nebulae, and stars. Australia makes of 20° north. Wrapped on three sides by the constellation Centaurus, it includes several notable stars in its bright "t": Alpha Crucis, or Acrux, a double star at the foot of the cross; Gamma Crucis, or Gacrux, a double star at the northern end; and Beta Crucis, or Mimosa, which ends the cross's

Crux (center) and the Coalsack Nebula slightly to its left

a fine sky watching site, with its clear, dry, dark skies. In April or May, Australia's autumn, the center of the Milky Way passes overhead, dense with stars. Below are listed just a few of the southern sky's finest sights.

Crux

A small but distinct constellation, the Southern Cross, or more formally, Crux, is best seen from latitudes south left arm. A gorgeous open cluster, the Jewel Box, gathers around the bright golden star Kappa Crucis, just below Beta Crucis. Through binoculars, you can see it sparkle in different colors.

Within Crux is a conspicuous dark patch just east of Alpha Crucis. Called the Coalsack Nebula, this is a dense cloud of gas and dust obscuring the stars behind it. It is easily seen silhouetted against the Milky Way.

Centaurus A galaxy

Centaurus

This huge constellation is said to represent the centaur Chiron, who tutored Hercules. Located south of the constellation Hydra, it contains a wealth of notable objects. Among them are its brightest star, Alpha Centauri. This binary star (with Beta Centauri) is one of the Sun's nearest neighbors, only 4.3 light-years away. A nearby companion visible only in telescopes, Proxima Centauri, is even closer, some 4.2 light-years distant. Centaurus also hosts a spectacular globular cluster, Omega Centauri (NGC 5139). Containing more than a million stars, this brilliant cluster is easily visible to the naked eye.

Large & Small Magellanic Clouds

These wispy glowing patches are irregular dwarf galaxies that orbit the Milky Way. The Large Magellanic Cloud (LMC) is found in the constellation Dorado, not far from the south celestial pole; the Small Magellanic Cloud (SMC) floats within the constellation Tucana. The LMC is 168,000 light-years distant, while the SMC is about 200,000 light-years away, both of them much closer to Earth than the Andromeda galaxy.

Crux's four stars (lower center)

Comets & Meteors

9

Comet Hale-Bopp streaks through the sky in 1997.

Comets

Ion Tail

Dust Tail

Coma

To Sun   Nucleus

Comet Hale-Bopp

A S THE FLAT PLANE of the new solar system took shape, billions of balls of frozen gas, laced with dust, also formed. Many of these "snowballs" are thought to have been swept by the gravitational forces of Jupiter and Saturn out to the distant Oort cloud. Others formed just outside the orbit of Neptune, creating what is now known as the Kuiper belt and a far-flung overlapping area called the scattered disk. We see these ancient objects when they are nudged into a new orbit and soar through the inner solar system as comets.

Comet Composition

A stew of frozen chemical compounds, comets begin to warm up as they approach the Sun and release the gases stored during the solar system's early days. The gases form a glowing head, or coma, around the frozen nucleus. A flow of charged ions is shaped by the solar wind into a glowing gas tail.

SKYFACT

King Louis XV dubbed French astronomer Charles Messier the "Comet Ferret" for his constant discovery of new ones.

Meanwhile, a stream of dust, also shed by the nucleus, forms a second tail.

Once in an orbit that will pass through the inner solar system, comets tend to follow an elliptical path. Those dislodged from the Oort cloud become long-period comets or get slung from the solar system. Those that originate in the Kuiper belt or scattered disk evolve into short-period comets. Comets typically are large enough to survive perhaps hundreds of trips through the inner solar system.

Observation

You can observe comets known to be coming or hunt for new ones. About two dozen are discovered each year, a few bright enough to be spotted by amateur equipment. If you see something not on a chart, make sure it is not a star, and look for a tail.

Recheck it to see if it has moved. Make notes on the location and magnitude, sketch the appearance, and have someone confirm what you've seen. Also, make sure it's not a known object. If it seems to be a comet, you can notify the Central Bureau for Astronomical Telegrams in Cambridge, Massachusetts.

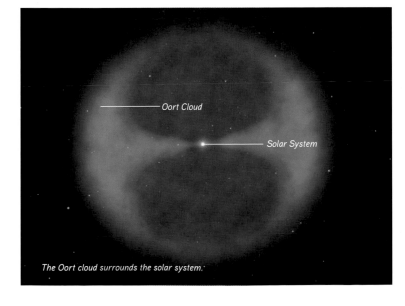

The Oort cloud surrounds the solar system.

Oort Cloud

Solar System

Famous Visitors

I N THE LATE 1600s, with Isaac Newton's math-based view of the world just beginning to take hold, astronomer Edmond Halley tried to use those new ideas about gravity and motion to see if he could predict—as Newton's theories seemed to allow— the return of any of the comets noted in historical records.

Establishing tables of comet observations, Halley noted that one of these wandering bodies had turned up in 1531, 1607, and 1682—and it was headed in roughly the same direction each time. He ventured a bold guess: The three comets were the same object, and it would appear again in late 1758—a prediction that proved to be accurate.

Named in his honor, even though he never saw it, the object that is known as Halley's comet has been mentioned as far back as the second century B.C. in cuneiform tables, was stitched into the famous Bayeux Tapestry depicting the Battle of Hasting in 1066, and is still regarded as an event in modern times.

Halley's comet had its last swing by Earth in 1986. It won't be back until 2061, but the attention given to the comet speaks to the primal nature of celestial phenomena. When something new appears in the sky, people

Comet Halley's most recent visit in 1986

notice—either out of fear, wonder, or a combination of both.

Comet Sightings

Comet Kohoutek in 1973 was a widely publicized bust that—despite the fact that it was visible to the naked eye—did not seem to match the "comet of the century" billing it received when it was discovered.

The truly bright ones don't happen very often. But scientists and sky watchers still have plenty to work with, particularly in recent years. Three bright comets put on a show in the 1990s. Comet Hale-Bopp in 1997 and comet Hyakutake in 1996 were watched by professionals and amateurs alike when they arrived—but both are long-period and won't be back soon.

Comet Shoemaker-Levy 9 in 1994 earned special attention when it was captured and ripped apart by Jupiter's gravity. The col-

Edmond Halley

lision of its shredded pieces with the massive planet caused a series of spectacular explosions. The encounter and its aftermath were studied closely for clues about the effects of such interplanetary calamities.

Comet McNaught—the "Great Comet of 2007"—was the brightest visitor in many decades but it is on an apparent path to leave the solar system and never return.

The most recurrent visitor is comet Encke, which cycles by every 3.3 years. Although it is little more than a fuzzy patch—when it is even visible to the naked eye—it is perhaps the most studied comet of them all. Detected first by French astronomer Pierre Mechain in 1786, it has been tracked through dozens of visits since then. Theories about comet Encke cite it as the possible source of major planetary impacts in the past, possibly when a larger object broke into pieces.

THE SCIENCE OF
Missions to Comets

The European Space Agency's Giotto spacecraft in 1986 analyzed the composition of Halley's comet; its Rosetta mission is intended to place a lander on comet 67P/Churyumov-Gerasimenko. NASA's Stardust spacecraft provided samples of comet Wild 2 for analysis, leading to new theories about comet formation. Its Deep Impact probe in 2005 provided evidence of water ice and organic matter on comet Tempel 1.

Meteors

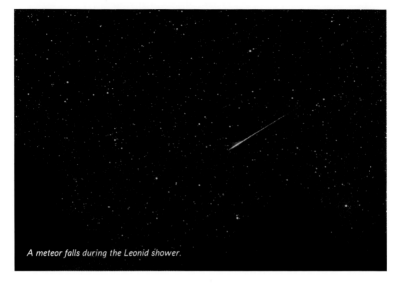

A meteor falls during the Leonid shower.

METEORS—shooting stars—are visible almost any night to the patient observer.

Meteoroids

Meteors begin as meteoroids—interstellar dust and debris. Individual particles are typically small, though some objects may be several feet wide. At that larger scale, the distinction between meteoroids and asteroids is somewhat arbitrary (the International Astronomical Union defines meteoroids as "considerably smaller than an asteroid and considerably larger than an atom or molecule"). The larger ones are thought to be pieces of asteroids, planets, or our Moon that have been broken off and set adrift.

But most meteoroids are small grains of interstellar material, often the "ash" burned from the outer layers of comets and left in massive trails around the solar system. As Earth progresses in its orbit, the planet encounters upwards of half a billion of these tiny objects every year.

Earthly Contact

These small bodies become meteors when they enter Earth's atmosphere. Under its gravitational pull they reach speeds of anywhere from 20,000 to more than 160,000 miles (32,000 to

SKYFACT

Sixty-four or so meteorites are known to be pieces from the Moon, and 34 are fragments of Mars. There may well also be meteorites on Earth from Venus, Mercury, and other bodies in the solar system, but none have yet been specifically identified.

Meteorite discovered in 1836

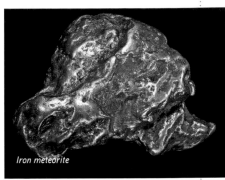

Iron meteorite

more than 257,000 kilometers) an hour. Friction from the atmosphere heats them to perhaps 2000°F (1093°C), a process that creates a glowing streak across the sky, which is why they are called shooting stars. Under such extreme heat and pressure, most meteors are vaporized before they get within 50 miles (80 kilometers) or so of Earth. Most are so small that they last barely a second before destruction, and may be so quick and dim that they cannot be seen from Earth. Larger ones create long, bright streaks that leave trails of smoke in their wake, or break up in a flash or two of light, causing what is called a bolide, although they are rare.

Observation

On any given night, there may be four or five visible shooting stars each hour. Annual showers occur when the Earth passes through streams of dusty debris from comets. Several dozen meteors may be visible each hour, emerging from what is called the shower radiant. The constellation where the radiant lies lends its name to the shower.

Occasionally, pieces of meteor reach the Earth—meteorites. The biggest ones, properly asteroids, have caused mass extinctions. The most prevalent are made mostly of silicates and other rocky material; some appear to have been formed from the same early solar system matter that produced the Sun. About 6 percent are iron meteorites, composed of iron and nickel. A small number of stony-iron meteorites are a roughly equal mixture of the two types of material.

THE STORY OF Sacred Stones

Like comets and other unfamiliar lights in the sky, meteors have been viewed as portentous. Those that actually reach Earth have taken on even more importance. A meteorite is thought to have been the source of a stone placed at the Temple of Delphi and venerated by the ancient Greeks. Around A.D. 220, the Roman emperor Bassianus made the Stone of Emesa—most likely a meteor—an object of public worship. Some theorize that iron-nickel alloys from meteorites provided the first usable source of the metal in the early Iron Age, and provided clues about how to manufacture it from ore.

Falling to Earth

THE FACE OF the Moon tells the story of the solar system's violent early years. Our planet has been bombarded plenty as well, but while the Moon's stable geology has maintained evidence of its bombardment by asteroids and meteoroids, Earth's tectonic plates, vast sea cover, and changing surface have hidden or erased most dents and dings.

Crashes & Craters

The largest recent event occurred in 1908, in a remote part of Russia, when an explosion decimated 830 square miles (2,150 square kilometers) of mostly uninhabited land in a blast that unleashed the equivalent of 15 million tons of TNT. The Tunguska event most likely involved a roughly 100,000-ton comet or asteroid exploding within about six miles (ten kilometers) above Earth's surface.

No impact crater was found, but researchers have identified 175 craters and larger impact basins worldwide associated with comets, asteroids, and meteoroids. Some, such as in Sudbury, Canada, carried enough metal to help local industries and enrich surrounding farmland. With a width of 0.75 mile (1.2 kilometers) and a depth of 600 feet (183 meters), the 50,000-year-old Barringer Meteor Crater, located between Flagstaff and Winslow, Arizona, resembles the lunar surface.

SKYFACT Explorer Robert E. Peary deduced the source of iron tools among Greenland Inuit when a local guide helped him track down a meteor split into three deposits, including a 34-ton chunk. Peary brought them back to the U.S. and sold them to the American Museum of Natural History in New York.

Barringer Meteor Crater, Arizona

The Chicxulub impact may have caused a mass extinction 65 million years ago.

The 110-mile (177-kilometer) Chicxulub crater, found on Mexico's Yucatán Peninsula in the 1970s, has been buried by sediment, but exploration has shown the geologic markings of an asteroid strike—glassy tektite rocks and large amounts of iridium, among other things. The object is estimated to have been 6 to 12 miles (10 to 19 kilometers) wide, and has been dated to the end of the Cretaceous period, consistent with the disappearance of many Earth species, suggesting that the asteroid's dust blocked sunlight, killed much of the plant life, and caused temperatures to plummet.

THE SCIENCE OF Impact Odds

By some estimates, a collision large enough to cause mass extinction will occur roughly every 100 million years. Objects big enough to destroy a city are estimated to hit perhaps every thousand. NASA is scanning known asteroids to assess the odds; on the 0 to 10 Torino scale—with 10 being a 100 percent chance of catastrophic impact and 0 being no chance—only 1 of 217 objects rates more than 0, and most of them rate 1 before being quickly downgraded. On October 6, 2008, scientists noticed an object, 2008 TC3, and calculated that it was on a collision course with Earth. It entered the atmosphere over Sudan the next day, releasing the equivalent of up to 2,000 tons of dynamite but incinerating before it hit the ground.

Meteor Showers

THE OCCASIONAL shooting star that enlivens a campout or a stroll on the beach is called a sporadic meteor, and on a typical night perhaps a half dozen or so can be spotted every hour. However, about 30 times a year Earth passes through a dense dust trail, typically left by a comet, giving rise to a meteor shower or an even more spectacular (if rare) meteor storm.

As comets move through the solar system, the outer layers are vaporized and emit billions of tons of material in the wake of their orbit. When Earth encounters or passes near one of these dust patches, the results can be spectacular. The seemingly random loca-

The Leonids occur every year in November.

tion of sporadic meteors is replaced by a dense and steady stream of shooting stars emanating from a specific part of the sky. These encounters occur about the same time each year, and even at their lightest might produce a couple dozen meteors every hour. Heavy storms might produce 10 or 20 shooting stars a second.

Forecasting Showers

Since meteor storms occur with regularity, their appearances are well publicized—in astronomy magazines, by organizations like the International Meteor Organization, and probably by your local meteorologist as part of the weather report. There is no shortage of information, in other words, about when the annual storms are expected to occur and peak.

Observation

A few preparations will make for more successful viewing. Foremost, be conscious of where the particular storm is originating—a location known as its radiant—and when visibility in that particular part of the sky is best from your location. The radiants are typically identified with constellations. If Leo is near the horizon in mid-November, when the annual Leonid (children of Leo) shower occurs, that will require finding an adequate viewing spot. You may find out that the storm is peaking during daylight hours, giving you the choice of accepting a less than maximum view at night and early in the morning, or opting out until next year. In that case, predawn trips are usually

better; they put you in a position facing the planet's path on its orbit, heading into the dust trail.

Give yourself time; using a lounge chair can save your neck from hours of looking up. Estimates are often given about how many meteors a storm may produce in an hour, technically known as the zenithal hourly rate. But that figure is premised on viewing under dark conditions when the radiant of the storm is overhead. If, instead, the radiant is low and you are viewing the shower at dusk, under a full Moon, or near city lights, the number will drop. No equipment is needed, but scanning the relevant area with binoculars may help locate some of the fainter shooting stars.

And don't ignore the rest of the sky. Meteors may emerge anywhere during a shower. Take a friend along, and alternate focusing on the radiant and looking elsewhere.

253

Annual Showers

ANNUAL METEOR showers can vary widely in their intensity, but a handful are among the more consistent and generally worth a look, both for serious hobbyists and casual naked-eye observers.

Winter Storms

The Geminids fall between December 7 and December 17, with a peak around December 13. Producing perhaps a hundred meteors an hour, the Geminid shower is one of the densest annual showers, and also one of the best to view in the evening (as opposed to after midnight), since the dust trail meets Earth's orbit opposite the Sun. The source of the meteors is object 3200 Phaethon, an asteroid that may well be an "extinct" comet nucleus, one whose ices have become exhausted or deeply buried by surface dust.

SKYFACT

The intensity of meteor showers is related to how soon after a comet's passage Earth intercepts the dust trail.

The November 1833 Leonid meteor storm

Major Meteor Showers

SHOWER	DATES OF ACTIVITY	PEAK
Quadrantids	Jan. 1–5	Jan. 3
Eta Aquarids	Apr. 19–May 28	May 4
Perseids	July 17–Aug. 24	Aug. 12
Orionids	Sept. 10–Oct. 26	Oct. 22
Leonids	Nov. 14–21	Nov. 17
Geminids	Dec. 7–17	Dec. 13

Best viewed in the farther northern latitudes, the Quadrantids flow from the constellation Boötes and produce up to a hundred shooting stars an hour. The shower is a New Year's event, running from January 1 through January 5 and peaking around January 3. Its "parent" was only recently identified as near-Earth asteroid 2003 EH1.

Spring & Summer

Halley's comet left a dust trail Earth approaches twice a year, producing two meteor showers, including one major one: The Eta Aquarids which take place from April 19 to May 28, with a peak around May 4. Originating from the middle of the constellation Aquarius, the shower produces up to 20 meteors an hour, though more than double that can be seen in the Southern Hemisphere.

Originating in the constellation Perseus, the summertime Perseids run from around July 17 to August 24, with a peak around August 12. Meteors vary in their apparent speed depending on whether they are catching up with Earth or meeting us head-on. The Perseids are swift, traveling perhaps 133,000 miles (214,000 kilometers) an hour. The flow of shooting stars can reach as many as 80 an hour. The Perseids were noted in Chinese records dating to A.D. 36. They were also the first meteor shower identified with a particular comet when astronomer Giovanni Schiaparelli in the 1860s identified it with the newly discovered comet 109P/Swift-Tuttle.

Autumn Showers

Emanating from Orion, the Orionids occur from September 10 to October 26, peaking on October 22 at around 25 meteors an hour. The Leonids, emanating from Leo, have been the source of repeated intense meteor storms on a 33-year cycle, the orbital period of its parent comet, 55P/Tempel-Tuttle. But such peaks are becoming highly variable. The Leonids occur from November 14 through November 21, with a peak on November 17. With only 10 to 15 meteors an hour, perhaps half of the meteors will leave a long-lasting train across the sky, like the vapor trail of a jet. They are the fastest moving meteors, speeding head-on toward Earth at 45 miles (72 kilometers) a second.

THINKING OF investing in a telescope to view the heavens? Before you do, here are some things to consider. First, choosing a telescope should be done carefully. Newcomers might lean toward telescopes advertised as starter instruments for astronomy. These usually feature a long white tube, tall skinny legs, and a price tag around $100 or less. And invariably, after a few nights of frustrating use, these go straight to the back of the closet.

Places to Purchase

As with hobbies like photography, music, and skiing, astronomy is a leisure activity where bargains are comparatively scarce. Beginners should avoid trying to buy in at the lowest prices advertised. In the long run, it will pay to pick up some knowledge and to budget realistically before pulling out your wallet.

If you're serious about exploring astronomy, two places to avoid are toy stores and "big box" stores. A telescope sold in a toy store is likely to be just what you'd expect: a toy. A telescope from a "big box" store is also likely to be of low quality, mass-produced, and aimed at people who will use the scope only once or twice before putting it away forever—or at least until the next yard sale. These instruments have soured more people on backyard astronomy than they have ever delighted.

The best place to start is a specialized store, where experienced staff

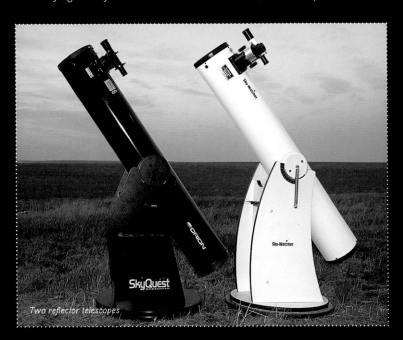

Two reflector telescopes

Large Newtonian reflector

members can help you find the right telescope for your wants and needs. The telescope you get won't be as inexpensive as in the other stores, but chances are you'll see a greater return on investment since you are more likely to be using it for years to come.

Consider Binoculars

Before making the jump to a telescope, consider binoculars as a first choice. Binoculars allow users to see far more stars and objects than they could with their eyes alone, and, at the same time, offer a wide field of view to appreciate some exotic celestial sights—like nebulae, clusters, and comets. Long-experienced observers regularly pack binoculars as part of their sky watching kit. Binoculars are also a good way to get a closer look at the Moon, and even some planetary delights, like Jupiter's largest moons, will show themselves in a good pair of binoculars. Plus, you can take them to sports events (which is hard to do with a good telescope).

As with telescopes, the best strategy in selecting a pair of binoculars is to choose quality over cheapness: The approximate minimum price of decent binoculars for astronomy purposes is in the range of $150. Try to avoid zooms, and keep the magnification to 10x or less. If your binoculars feature more than 10x magnification, you may find you have a hard time keeping the image steady in your eyepiece unless you are using a tripod.

Then look for light-gathering ability: how much light gets to your eyes. That's related to the size of the binocular's objective lens (the big lens at the front). The diameter of the objective lens in millimeters divided by the magnification power will give you the "exit pupil"—the size of the image coming out of the binoculars. For example, a 7x50 pair of binoculars will give you an exit pupil of 7.1 mm. (For 10x50s, the exit pupil is 5 mm.) Seven millimeters is close to an ideal size because the measurement matches the maximum size of a dilated human pupil.

Schmidt-Cassegrain telescope

Types of Telescope

I NSTRUMENTS FOR amateur sky watchers fall into three main types: refractors, reflectors, and catadioptric telescopes. Each one has its own advantages, and consideration should be given to how and when you will be using your telescope.

Different Types

Refractors look like what most people envision when you say the word, "telescope." These kinds of instruments feature an objective lens at the front end of the telescope tube, which collects light and directs it toward an eye-

Catadioptric telescope

Refractor in the field

piece at the other end of the tube. The eyepiece then magnifies the image.

Reflectors do away with the objective lens and instead use a parabolic mirror at the bottom of the telescope tube. Incoming light falls down the tube onto the mirror, is reflected upward toward a secondary mirror, which directs the light through the side of the tube to the eyepiece, which magnifies the image.

Catadioptric telescopes combine a corrector lens with a mirror. Light enters the tube through the transparent corrector lens, bounces off a main mirror at the back of the tube, goes up to a secondary mirror and reflects back down the tube, where it passes through a central hole in the main mirror to the eyepiece which lies behind.

Pros & Cons

Each type of telescope comes with pluses and minuses. Refractors, for example, tend to have long focal ratios (which is determined by dividing the size of the objective lens or mirror by the distance it takes that lens or mirror to form an image). Higher focal ratios

offer more sharply contrasted images, while lower focal ratios—such as those you'll typically find in reflectors, allow more sensitivity to light and wider fields of view.

If your main interests are to look at the Moon, the planets, double stars, or other such high-magnification phenomena, you might be best served with a refracting telescope. If, on the other hand, you are more interested in seeing star clusters, comets, and nebulae, a reflector might be a better fit for your needs. If you're looking to see a bit of everything and want an all-purpose telescope, a catadioptric telescope, whose focal ratios are generally between those of reflectors and refractors, might be the best bet.

Reflector on a tripod

Using Your Telescope

Putting your new telescope to use can be rewarding if you've considered the right things before purchasing. A few practical considerations can make your experience that much better.

Size

Before you buy a big telescope, think about how and when you want to use it. Size matters—but for different reasons than you might expect. With telescopes, as in many things, there's often an unstated belief that bigger is better. To a limited extent this may be true. The larger your telescope's objective lens or mirror, the more light will enter it and the more objects are visible through your scope.

But that increase in size comes at a cost—first, a literal cost, as the bigger a telescope gets, the more expensive it becomes. Second, there's the hassle of dragging out a larger scope, which can weigh hundreds of pounds and

can be unwieldly to set up and use. If a smaller, lighter telescope makes it easier for you to examine the night sky on a whim, choose one of those. Also something to consider: A 4-inch telescope will typically allow you to see objects down to the 12th magnitude, which is more than enough to see all the planets in our solar system and any number of nebulae, clusters, and galaxies.

Eyepieces

Eyepieces are high-quality magnifying glasses you use to examine the image formed by the telescope's main optics. Changing the eyepieces changes the telescope's magnification. Some telescope deals offer a slew of eyepieces yielding high magnifications, but don't be taken in: The reality is that few nights have steady enough skies to let you use more than about 35x per inch of telescope aperture. (On a 6-inch scope this would be 210x.)

Computerized mounts can help track an object.

Sharing the night sky at a star party

Most of your observing will be done with just two or three eyepieces: probably ones that yield around 50x and 150x and occasionally 300x. If the eyepieces that come with your telescope are of low quality, you may need to upgrade. When you do, consult a salesperson to make sure your eyepiece is appropriate both for the type and size of your telescope. In addition to eyepieces, telescopes can be fitted with a wide range of filters, from polarized filters to help resolve details on bright objects like the Moon, to colored filters that can bring out detail highlights on planet surfaces.

Computers

Many new telescopes come with computerized mounts. These help skywatchers keep their favorite sights in view as they move across the sky (or more accurately, as the earth rotates away from them). They can also help stargazers locate interesting sights in the sky simply by punching in coordinates or even just the name of the object they want to look at.

Some sky watching purists feel software takes some of the challenge out of astronomy, but most find it useful because it quickly locates the best the sky has to offer. One thing to keep in mind, however, is that even if you buy a computerized telescope, you will still likely need to locate a guide star—a bright star you know—so the telescope can orient itself.

Togetherness

Local astronomical societies often get together on a regular basis to share their enthusiasm for astronomy and the night sky. The membership is a great source of information on telescopes, technical troubleshooting, and everything that's going on in the heavens. You will learn a lot—and will very likely get more out of your telescope and the time you spend using it.

Deep Space

Billowing clouds of gas make up the Carina Nebula.

The Universe

THE STORIES ancient cultures used to explain the cosmos sound strange today. But are they any stranger than the reality? Consider: An infinitely small point of infinite density and gravity erupts to create not just space but time, unleashing a torrent of radiation that cools, starts to form matter, which starts to form stars, galaxies, planets—us.

Fundamental questions remain unanswered (and perhaps unanswerable), but recent information has cre-ated a more detailed picture of how the universe evolved.

Age of the Universe

The clock can't yet be pushed back to "time zero," but recent data from the Wilson Microwave Anisotropy Probe (WMAP) have led to what scientists regard as an accurate estimate of the universe's age. The WMAP, launched in 2001 into an orbit more than 1,000,000 miles (1,600,000 kilometers) from Earth, developed a

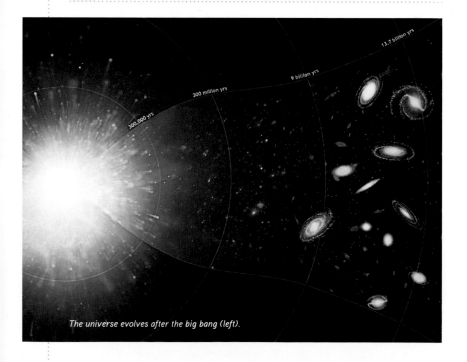

300,000 yrs

300 million yrs

9 billion yrs

13.7 billion yrs

The universe evolves after the big bang (left).

Matter as we see it around us is now thought to make up only about 4.6 percent of the universe. Using the temperature mapping from WMAP and other data, scientists estimate that about 23 percent of the universe consists of cold dark matter—objects whose presence is witnessed in the fact that light from distant galaxies bends in a way that the intervening stars don't explain. That galaxies hold together, even though the mass of visible matter is not adequate to keep them from ripping apart, is also evidence of dark matter. The remaining roughly 72 percent of the universe is made of dark energy, whose existence, so far, can only be inferred. Observations of supernovae in the 1990s indicated that the universe is expanding and that the expansion rate is accelerating—a phenomenon explained by an energy source countering the effects of gravity.

detailed map of cosmic background radiation and is measuring differences in the temperature of that background radiation across the sky. Though minuscule, the variance explains how heavier elements, stars, and galaxies arose out of an early cloud of radiation and matter assumed to be evenly distributed. The temperature map created by the WMAP is consistent with the emergence, over 13.7 billion years, of a universe that looks like ours. Based on estimates from the Hubble Space Telescope, that universe contains perhaps 120 billion galaxies, each containing into the billions of stars.

More Questions

The age of the universe is only one riddle. It seems to be expanding faster than earlier theories would predict, leading to the premise that repulsive "dark energy" is forcing galaxies away from each other. By comparing the amount of estimated visible matter in a galaxy with the gravitational force needed to hold it together, scientists

have concluded there's something missing. So-called dark matter supplies that missing mass: Brown dwarf stars, black holes, and other massive compact halo objects (MACHOs) play a role, as may still hypothetical entities known as weakly interacting massive particles (WIMPs).

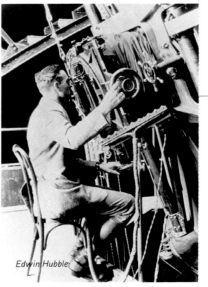

Edwin Hubble

Galaxies

OUR MILKY WAY and other galaxies are made of stars, interstellar dust, gas, and dark matter (not fully understood, but including an assortment of objects like brown dwarf stars, planets, black holes, and some more exotic objects that do not emit light). They come in three basic shapes—elliptical, spiral, and irregular.

Galactic Shapes

Elliptical galaxies, which make up about 18 percent of all galaxies, may be near spherical, or be stretched into an oblong. Stars in elliptical galaxies, which are composed mostly of older, red giant stars, follow their own paths around the galactic center.

Spiral galaxies rotate around a bright nucleus. Arms spiral out from that point. Spirals account for most galaxies: 78 percent of all. All spiral galaxies hold a mix of old and young stars, and typically have dusty regions of new star formation.

Irregular galaxies, 4 percent of the total, are amorphous collections of stars that include star-forming nebulae but lack any coherent shape.

Collisions

Galaxies are so massive that the strength of one can rip apart another,

SKYFACT

The Milky Way and Andromeda galaxies may have formed closer together, before a collision with a dwarf galaxy drove them apart.

See "Our Galaxy," pp. 128-129

The Sombrero galaxy (M104) lies in the constellation Virgo.

Elliptical galaxy

Spiral galaxy

Barred spiral galaxy

Irregular galaxy

while even in the relative emptiness of space they can collide. For example, in the constellation Corvus, the Antennae galaxies are in the process of merging together. In some cases large ellipticals may simply absorb nearby dwarf galaxies.

Clusters

Each containing billions of stars, galaxies are themselves arrayed in clusters around the universe. The Local Group, for example, includes the Milky Way and Andromeda as the dominant members, and around 30 much smaller neighbors scattered over a distance of about three million light-years. Other clusters have thousands of members. And the clusters themselves form into larger structures.

It seems that superclusters also have shape, grouping into large-scale structures that give the overall universe a "clumpy" nature. Galaxies, clusters, and superclusters seem to be arrayed in strands, much longer than they are thick.

THE SCIENCE OF
The Great Attractor

As if a black hole at the center of the Milky Way was not worrisome enough, the "Great Attractor" may have its own surprises. Located some 150 million light-years away, this aggregation of galaxies and dark matter is drawing in everything around it, including our Local Group of galaxies. It is moving in that direction at roughly 500 miles (805 kilometers) a second, as is the rest of the Virgo supercluster. Obscured at first by the Milky Way's gas and dust, a cluster of galaxies known as Abell 3627 has been pinpointed by space-based observation as the center of a massive supercluster causing the "attraction."

The Farthest Reaches

The Hubble telescope reveals the farthest reaches of space.

LIGHT HAS A FINITE speed—186,000 miles (299,000 kilometers) a second. The implications of this are easy to grasp on a "local" level. Since the Sun is about 93 million miles (150 million kilometers) from Earth, the light leaving its surface takes about 500 seconds to reach our eyes. In other words, we see the Sun as it was 8.3 minutes ago. Moonlight falling on the patio left the Moon 1.3 seconds ago.

Now carry that over a galactic or universal scale, and the results are a bit more mind-boggling. Train a backyard telescope on the Andromeda galaxy, at 2.5 million light-years away our nearest galactic neighbor, and you are literally peering back in time—and seeing that collection of stars at it was about the time *Homo habilis* emerged in Africa.

THE SCIENCE OF Quasars

A discrepancy in the early 1960s over the massive output of light of an object known as 3C 273 led to some efforts to redefine how the concept of redshift—a way of estimating distance—was used and the conclusion that quasars have at their heart a supermassive black hole causing a brilliant stream of energy to be released as it consumes the contents of nearby stars.

Peering Deeper

More sophisticated equipment goes even deeper into space and further back in time. The Hubble Space Telescope has gone the furthest, collecting information from more than 12 billion light-years away. In a universe calculated to be 13.7 billion years old, Hubble has provided information about galaxies and stars as they existed near the beginning of time.

That has helped astronomers develop a picture of how the universe and its galaxies evolved. Among the questions that such observation has helped resolve are the nature of quasars and their relationship to supermassive black holes. Quasars were first detected in the 1960s from their radio emissions. At first, there was no visible object attached to the signals that radio astronomers had identified. When the source was found, it appeared to be a normal star, yielding the newly coined term "quasar," for quasi-stellar radio source.

Distant Objects

The more the new objects were studied, however, the stranger they seemed. For example, calculations using the redshift of light from quasars meant that they were among the universe's most distant objects—far too distant, in fact, for light from ordinary stellar fusion to reach Earth at the magnitude that the quasars were producing. Theories about them included conjecture that they were powered by black holes whose interaction with surrounding stars and gases created light emissions perhaps a trillion times as bright as the Sun.

Amid continuing controversy over quasars, the Hubble telescope helped clarify the situation in the 1990s. Its series of Deep Field images showed the stars surrounding quasar galaxies, leading astronomers to hypothesize that these intense energy sources are part of normal galactic development. Among the universe's more ancient objects, they were violently active during its early years, when collision between galaxies fed energy into them. Over time, the flow of energy decreased, and they evolved into the "normal" galaxies seen today.

A quasar

The Big Bang

S IR FRED HOYLE'S invention of the term "big bang" was certainly unintentional. The English physicist championed a "steady state" cosmology, in which matter was constantly being generated to fill an expanding universe—a stable process that could go on indefinitely. He coined big bang out of sarcasm, to deride those who felt the universe began in a single cataclysm. But the name stuck. Hoyle's insult now refers to a theory that, while hardly perfected, has been consistent with what is detected as we peer deeper into space.

Universal Origins

If current thinking is correct, the universe began in a "singularity," a point of infinite gravity, where space,

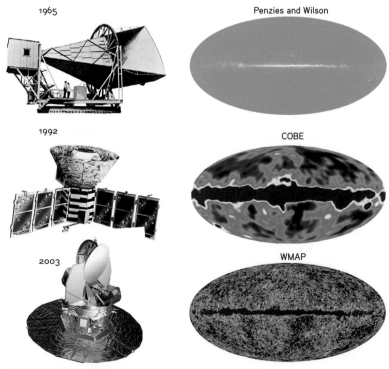

1965   Penzies and Wilson

1992   COBE

2003   WMAP

Satellites reveal more background radiation from the early universe.

Backyard Guide to the Night Sky

time, and all subsequent matter and energy were contracted into an object without size. The concept has led theoretical physicists to try to explain how such entities could exist, yielding to exotic constructions like string theory—the idea that everything is made of massless filaments of energy. Those filaments would exist in more dimensions than the four (three in space and one in time) that current theories explain.

Current cosmology has not yet pushed back to that moment of origin but does try to explain the chain of events that unfolded less than a second afterward. In this beginning, there were gamma rays—high-temperature, high-energy photons. As the cataclysm unfolded and space-time expanded, the gamma rays cooled and decayed into basic particles of matter. Shortly after, the simplest atoms—hydrogen and helium—had formed in the roughly three-to-one ratio found in the most ancient observed stars.

Initially (and in this case, that's perhaps a million years) these elements existed within a fog of radiation. At about the 300,000-year mark, however, free atomic particles were increasingly absorbed into stable atoms. It is at this point that the universe became "transparent," allowing devices like the Wilson Microwave Anisotropy Probe to measure the radiation left from the initial event.

The existence of the cosmic background radiation marked a turning point in the debate over the universe's origins. Trying to determine the source

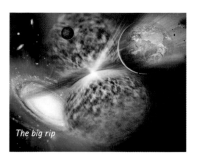

The big rip

of static that was corrupting satellite signals, Bell Labs engineers Arno A. Penzias and Robert W. Wilson in 1965 famously determined that the "noise" was coming from the entire sky—at a wavelength consistent with background radiation from the big bang.

THE SCIENCE OF The Universe's Ultimate Fate

Like the big bang, the long-term fate of the universe will certainly be big—a big crunch, big rip, or big bounce. Momentum from the universe's formation led to its expansion. At the same time, gravity works to slow the process down. Depending on the density and pressure of matter in the universe, gravity might win the battle, causing the universe to contract—either back into a singularity (the big crunch), or only down to the point where heat and rising pressure cause it to expand again (the big bounce). Or gravity may lose and the expansion continue. That is where the evidence is pointing. Recent analysis from the WMAP produced the unexpected conclusion that the universe's rate of expansion is accelerating rather than slowing under gravity's influence, a fact attributed to the force of dark energy. Perhaps that expansion can continue forever, or alternatively will lead to a catastrophic big rip.

Looking Into the Depths

ABD AL-RAHMAN AL-SUFI, the tenth-century Persian astronomer, showed that even the naked eye can carry us beyond the Milky Way. His work showed what he deemed "the little cloud" in the constellation Andromeda—what we now know to be the Andromeda galaxy, 2.5 million light-years away, the closest galaxy to the Milky Way. Even exotica like quasars are within reach of the skilled amateur. Just five degrees northwest from what would be the Virgin's head in the constellation Virgo (the star Porrima, or Gamma Virginis), the quasar 3C 273 can be spotted some three billion light-years away through an 8-inch-aperture telescope.

Seeing Deep Space

Going deep into the sky requires practice and a willingness to use all

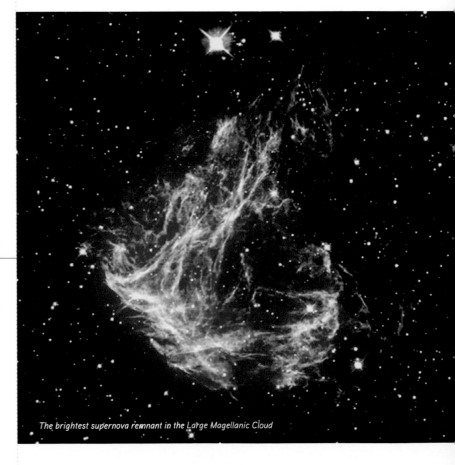

The brightest supernova remnant in the Large Magellanic Cloud

Backyard Guide to the Night Sky

An ordained Anglican minister, Reverend Thomas William Webb (1807-1885) had an affinity for backyard astronomy and pursued his interest with great passion and attention to detail. On the grounds of his parsonage in Hardwick, England, he built a small observatory where he used several different kinds of telescopes to observe the night sky. Based on his observations at Hardwick, in 1859 he wrote *Celestial Objects for Common Telescopes*, a guidebook for amateur astronomers. His book described what could be seen with a common telescope and contained directions on how to use different telescopes. The book remained the go-to guide for amateurs well into the 20th century, and Webb became considered the father of amateur astronomy.

the tools at hand. Clear, dark skies are a necessity: Even under ideal conditions, distant galaxies just barely outshine the background glow of the night (remember, the naked-eye stars themselves provide a baseline amount of light that any deep sky object must compete against). Give your eyes time to adapt to the dark—a minimum of 15 to 20 minutes. Get familiar with averted vision, in which you look slightly to one side of an object that hovers on the edge of detectability. This lets you use a part of the retina that can register more detail in dark conditions. Keep in mind the trade-off between magnification and field of view in the choice of a telescope

eyepiece: Spotting large galaxies requires lower power, a wider field of view, and a reliance on contrasts; smaller, brighter objects can be seen under greater magnification, provided you can capture it in the narrower field. Focus on the objects that are higher overhead. Light from objects closer to the horizon travels farther through the atmosphere to reach your eye and loses intensity.

Amazing Images

Although it won't carry the thrill of discovery or offer the chance to find a new comet, the most dramatic views of deep space will come through projects run by NASA, the European Space Agency, and astronomers at the major world observatories. They produce a steady flow of publicly available images and information about the studies being done and the conclusions being reached.

SKYFACT

Visit the Webb Society *(www. webbdeepsky.com)* to learn even more about the wonders of deep sky astronomy.

ECLIPSES
Be sure to consult astronomy magazines and other resources for exact viewing times. *All Eclipse Data by Fred Espenak, NASA's GSFC*

Total Solar Eclipses

Year	Date	Viewing Location
2009	July 22	India, Nepal, China
2010	July 11	Easter Island, Chile, Argentina
2012	November 13	Northern Australia
2015	March 20	Iceland, northern Europe
2016	March 9	Sumatra, Borneo, Sulawesi
2017	August 21	North America
2019	July 2	Chile, Argentina
2020	December 14	Chile, Argentina

Partial Solar Eclipses Visible from North America

Year	Date	Type	Viewing Location
2011	June 1	Partial	Canada
2012	May 20	Annular	Pacific northwest
2014	October 23	Partial	Western North America

Lunar Eclipses Visible from North America

Year	Date	Type	Location
2009	February 9	Partial	Western North America
2010	December 21	Total	North America
2011	December 10	Total	Western North America
2012	November 28	Partial	North America
2014	April 15	Total	North America
2014	October 8	Total	United States
2015	April 4	Total	Western North America
2015	September 28	Total	North America
2018	January 31	Total	Western North America

METEORS
Peak activity can vary, so consult resources for exact viewing times.

Major Annual Meteor Showers

Shower	Constellation	Dates of Activity	Peak
Quadrantids	Draco	Jan. 1–5	Jan. 3
Eta Aquarids	Aquarius	Apr. 19–May 28	May 4
Perseids	Perseus	July 17–Aug. 24	Aug. 12
Orionids	Orion	Sept. 10–Oct. 26	Oct. 22
Leonids	Leo	Nov. 14–21	Nov. 17
Geminids	Gemini	Dec. 7–17	Dec. 13

COMETS

Comet	Orbit Period	Next Date Visible
10P/Tempel 2	5.5 years	2009
81P/Wild 2	6.4 years	2010
1P/Halley	76 years	2061
109P/Swift-Tuttle	130 years	2127
C/1995 01 Hale-Bopp	2400 years	4397

VIEWING THE NAKED-EYE PLANETS
The naked-eye planets move throughout the sky during the entire year. The tables on pp.274-275 list the locations for the naked-eye planets through their annual journeys. With the exception of Mercury, the other planets will appear in constellations for months at a time during the year.

Mercury
Mercury can be found near the western horizon at dusk and the eastern horizon at dawn for a several two-week intervals throughout the year.

Year	At Dusk in the West	At Dawn in the East
2009	mid- to late April	early to mid-October
2010	late Mar. to early Apr.	mid- to late September
2011	mid- to late March	early September
2012	Early Mar.; mid-to late June	mid-August
2013	mid-Feb.; early to mid-June	late July and early Aug.; mid-Nov.
2014	late Jan to early Feb.; mid-May	late Oct. to early Nov.
2015	early May	mid-October
2016	mid- to late April	late Sept. to early Oct.
2017	late Mar. to early Apr.	mid-Sept.
2018	mid-March	Late August
2019	late Feb.; mid-June	mid-October

Transits of Mercury Across the Sun

Date	Universal Time	Separation*
November 8, 2016	14:57	319″
November 11, 2019	15:20	76″
November 13, 2031	8:54	572″
November 7, 2039	8:46	822″
May 7, 2049	14:24	512″

* Separation: distance (arc-seconds) between the centers of the Sun and Mercury. Transit Predictions by Fred Espenak, NASA's GSFC

Transits of Venus Across the Sun

Date	Universal Time	Separation*
June 6, 2012	1:28	553″
December 11, 2117	5:48	724″

* Separation: distance (arc-seconds) between the centers of the Sun and Mercury. Transit Predictions by Fred Espenak, NASA's GSFC

Venus

	Jan.	Feb.	Mar.	Apr.	May	June	July	Aug.	Sept.	Oct.	Nov.	Dec.
2009	Aqr	Psc	Psc	Psc	Psc	Psc	Tau	Gem	Cnc	Leo	Vir	Lib
2010	Sgr	Cap	Aqr	Ari	Tau	Gem	Leo	Vir	Vir	Lib	Vir	Vir
2011	Lib	Sgr	Cap	Aqr	Psc	Ari	Tau	Cnc	Leo	Vir	Lib	Sgr
2012	Cap	Aqr	Psc	Tau	Tau	Tau	Tau	Tau	Gem	Leo	Vir	Lib
2013	Oph	Cap	Aqr	Psc	Ari	Tau	Cnc	Leo	Vir	Lib	Sgr	Sgr
2014	Sgr	Sgr	Sgr	Cap	Psc	Ari	Tau	Gem	Leo	Vir	Lib	Oph
2015	Sgr	Aqr	Psc	Ari	Tau	Gem	Leo	Leo	Cnc	Leo	Leo	Vir
2016	Sco	Sgr	Cap	Psc	Ari	Tau	Gem	Leo	Vir	Lib	Oph	Sgr
2017	Aqr	Psc	Psc	Psc	Psc	Psc	Tau	Gem	Cnc	Leo	Vir	Lib
2018	Sgr	Cap	Aqr	Ari	Tau	Gem	Leo	Vir	Vir	Lib	Vir	Vir
2019	Lib	Sgr	Cap	Aqr	Psc	Ari	Tau	Cnc	Leo	Vir	Sco	Sgr
2020	Cap	Aqr	Psc	Tau	Tau	Tau	Tau	Tau	Gem	Leo	Vir	Lib

Mars

	Jan.	Feb.	Mar.	Apr.	May	June	July	Aug.	Sept.	Oct.	Nov.	Dec.
2009	Sgr	Sgr	Cap	Aqr	Cet	Ari	Ari	Tau	Gem	Gem	Cnc	Leo
2010	Leo	Cnc	Cnc	Cnc	Cnc	Leo	Leo	Vir	Vir	Lib	Sco	Oph
2011	Sgr	Cap	Aqr	Psc	Psc	Ari	Tau	Tau	Gem	Cnc	Leo	Leo
2012	Leo	Vir	Leo	Leo	Leo	Leo	Vir	Vir	Vir	Lib	Oph	Sgr
2013	Cap	Aqr	Aqr	Psc	Ari	Tau	Tau	Gem	Cnc	Leo	Leo	Vir
2014	Vir	Vir	Vir	Vir	Vir	Vir	Vir	Vir	Lib	Oph	Sgr	Sgr
2015	Cap	Aqr	Cet	Ari	Ari	Tau	Gem	Gem	Cnc	Leo	Vir	Vir
2016	Vir	Lib	Lib	Sco	Sco	Lib	Lib	Lib	Sco	Sgr	Sgr	Cap
2017	Aqr	Psc	Psc	Ari	Tau	Tau	Gem	Cnc	Leo	Leo	Vir	Vir
2018	Lib	Sco	Oph	Sgr	Sgr	Cap	Cap	Cap	Cap	Cap	Cap	Aqr
2019	Psc	Psc	Ari	Tau	Tau	Gem	Cnc	Leo	Leo	Vir	Vir	Lib
2020	Lib	Oph	Sgr	Cap	Cap	Aqr	Psc	Psc	Psc	Psc	Psc	Psc

Jupiter

	Jan.	Feb.	Mar.	Apr.	May	June	July	Aug.	Sept.	Oct.	Nov.	Dec.
2009	Sgr	Cap	Cap	Cap	Cap	Cap	Cap	Cap	Cap	Cap	Cap	Cap
2010	Cap	Aqr	Aqr	Aqr	Aqr	Psc	Psc	Psc	Psc	Psc	Aqr	Aqr
2011	Psc	Psc	Cet	Psc	Psc	Psc	Ari	Ari	Ari	Ari	Ari	Ari
2012	Psc	Ari	Ari	Tau	Tau	Tau	Tau	Tau	Tau	Tau	Tau	Tau
2013	Tau	Tau	Tau	Tau	Tau	Tau	Gem	Gem	Gem	Gem	Gem	Gem
2014	Gem	Gem	Gem	Gem	Gem	Gem	Gem	Cnc	Cnc	Cnc	Leo	Leo
2015	Leo	Leo	Cnc	Cnc	Cnc	Cnc	Leo	Leo	Leo	Leo	Leo	Leo
2016	Leo	Leo	Leo	Leo	Leo	Leo	Leo	Leo	Vir	Vir	Vir	Vir
2017	Vir	Vir	Vir	Vir	Vir	Vir	Vir	Vir	Vir	Vir	Vir	Lib
2018	Lib	Lib	Lib	Lib	Lib	Lib	Lib	Lib	Lib	Lib	Lib	Sco
2019	Oph	Oph	Oph	Oph	Oph	Oph	Oph	Oph	Oph	Oph	Oph	Sgr
2020	Sgr	Sgr	Sgr	Sgr	Sgr	Sgr	Sgr	Sgr	Sgr	Sgr	Sgr	Sgr

Saturn

	Jan.	Feb.	Mar.	Apr.	May	June	July	Aug.	Sept.	Oct.	Nov.	Dec.
2009	Leo	Leo	Leo	Leo	Leo	Leo	Leo	Leo	Leo	Vir	Vir	Vir
2010	Vir	Vir	Vir	Vir	Vir	Vir	Vir	Vir	Vir	Vir	Vir	Vir
2011	Vir	Vir	Vir	Vir	Vir	Vir	Vir	Vir	Vir	Vir	Vir	Vir
2012	Vir	Vir	Vir	Vir	Vir	Vir	Vir	Vir	Vir	Vir	Vir	Vir
2013	Lib	Lib	Lib	Lib	Lib	Vir	Vir	Vir	Lib	Lib	Lib	Lib
2014	Lib	Lib	Lib	Lib	Lib	Lib	Lib	Lib	Lib	Lib	Lib	Lib
2015	Lib	Sco	Sco	Sco	Sco	Lib	Lib	Lib	Lib	Lib	Sco	Oph
2016	Oph	Oph	Oph	Oph	Oph	Oph	Oph	Oph	Oph	Oph	Oph	Oph
2017	Oph	Oph	Oph	Sgr	Sgr	Oph	Oph	Oph	Oph	Oph	Oph	Sgr
2018	Sgr	Sgr	Sgr	Sgr	Sgr	Sgr	Sgr	Sgr	Sgr	Sgr	Sgr	Sgr
2019	Sgr	Sgr	Sgr	Sgr	Sgr	Sgr	Sgr	Sgr	Sgr	Sgr	Sgr	Sgr
2020	Sgr	Sgr	Sgr	Cap	Cap	Cap	Cap	Sgr	Sgr	Sgr	Sgr	Sgr

Further Resources

BOOKS

Aguilar, David A. *11 Planets: A New View of the Solar System*. Washington, D.C.: National Geographic Society, 2008.

Daniels, Patricia. *My First Pocket Guide: Constellations*. Washington, D.C.: National Geographic Society, 2002.

Dickinson, Terence. *Night Watch: A Practical Guide to Viewing the Universe*. Buffalo: Firefly Books, 2006.

Glover, Linda K. *National Geographic Encyclopedia of Space*. Washington, D.C.: National Geographic Society, 2005.

Lang, Kenneth R. *The Cambridge Guide to the Solar System*. Cambridge: Cambridge University Press, 2003.

Levy, David H. *A Nature Company Guide: Skywatching*. Alexandria, VA: Time Life Books, 1995.

WEBSITES

National & International Organizations
The American Meteor Society, *www.amsmeteors.org*
A non-profit website to encourage and support interest in Meteor Astronomy

British Astronomical Association, *britastro.org/baa*
Website for Amateur Astronomers in the United Kingdom

European Space Agency, *www.esa.com*
Official website for the European Space Agency (ESA)

International Astronomical Union, *www.iau.org*
Official website for the International Astronomical Union

National Aeronautics and Space Administration, *www.nasa.gov*
Official website for the National Aeronautics and Space Administration (NASA)

Space Weather Prediction Center, *www.swpc.noaa.gov*
Website run by the National Oceanic and Atmospheric Administration, that reports on the space environment around Earth

Periodicals
Astronomy, www.astronomy.com
Official website for *Astronomy* magazine

Sky & Telescope, *www.Skyandtelescope.com*
Official website for Sky & Telescope magazine

Observatories & Telescopes
Chandra X-Ray Observatory, *chandra.harvard.edu*
Official website for the Chandra, the first space-based X-ray observatory

European Southern Observatory, *www.eso.org/public*
Official website for the European Southern Observatory, for Astronomical Research in the Southern Hemisphere

Hubble Site, *hubblesite.org/newscenter*
The official website for the Hubble Space Telescope

National Optical Astronomy Observatory, *www.noao.edu*
Official website for Kitt Peak National Observatory, Cerro Tololo Inter-American Observatory, and National Optical Astronomy Observatory (NOAO) Gemini Science Center

Mauna Kea Observatories, *www.ifa.hawaii.edu/mko*
Official website for the world's largest observatory for infrared, optical, and submillimeter astronomy

Palomar Observataroy, *www.astro.caltech.edu/Palomar*
Official website for the Palomar Observatory in San Diego, California

W.M. Keck Observatory, *www.keckobservatory.org*
Website for the twin Keck Telescopes, the largest optical and infrared telescopes in the world

General Resources
Ask an Astronomer, *curious.astro.cornell.edu*
Site run by Cornell University where users can post questions for Cornell astronomers

Constellations, *www.astro.uiuc.edu/~kaler/sow/const.html*
The constellation website of Professor Emeritus Jim Kaler, of the University of Illinois at Urbana-Champaign

The Constellations and Their Stars,
www.astro.wisc.edu/~dolan/constellation
Website compiled by graduate student Christopher Dolan, of the University of
Wisconsin-Madison

Kids Astronomy, *www.kidsastronomy.com*
Free Astronomy website designed to teach
children about the Universe, run by Kidsknowit
Network

Moon Connection, *www.moonconnection.com*
Detailed website about the Moon, its phases,
and missions to the Moon

Nine Planets, *www.nineplanets.org*
A detailed overview of the planets, moons,
and other objects in our solar system

Space, *www.Space.com*
Multimedia website dedicated to space news

Space Weather, *www.spaceweather.com*
Information about the environment between
the Sun and Earth

NASA's Solar System Exploration,
solarsystem.nasa.gov
Dedicated to solar system facts and information

PLANETARIUMS & MUSEUMS
North- and southeast
Albert Einstein Planetarium
National Air and Space Museum
Independence Avenue at 6th Street, SW
Washington, DC 20560
www.nasm.si.edu/visit/theaters/planetarium

Hayden Planetarium
American Museum of Natural History
Central Park West and 79th Street
New York, New York 10024-5192
www.haydenplanetarium.org

Fels Planetarium
The Franklin Institute
222 North 20th Street
Philadelphia, Pennsylvania 19103
www2.fi.edu/theater/planetarium

Smithsonian National Air and Space Museum
Independence Avenue at 6th Street, SW
Washington, DC 20560
www.nasm.si.edu

Miami Science Museum
3280 S. Miami Ave
Miami, Florida 33129
www.miamisci.org

Midwest
Alder Planetarium
Adler Planetarium and Astronomy Museum
1300 S. Lake Shore Drive
Chicago, Illinois 60605-2403
www.adlerplanetarium.org

Museum of Science and Industry
57th Street and Lake Shore Drive
Chicago, Illinois 60637
www.msichicago.org

Nobel Planetarium
Fort Worth Museum of Science and History
1501 Montgomery Street
Fort Worth, Texas 76107
www.fwmuseum.org/noble

North- and southwest
Aerospace Museum of California
3200 Freedom Park Drive
McClellan, California 95652
www.aerospacemuseumofcalifornia.org

Dorrance Planetarium
Arizona Science Center
600 East Washington Street,
Phoenix, Arizona 85004
www.azscience.org planetarium

The Exploratorium
3601 Lyon Street
San Francisco, California 94123
www.exploratorium.edu

Oregon Museum of Science and Industry
1945 SE Water Ave
Portland, Oregon 97214
www.omsi.edu

Samuel Oschin Planetarium
Griffith Observatory
2800 East Observatory Road
Los Angeles, CA 90027
www.griffithobs.org

William Knox Holt Planetarium
University of California, Berkeley
Lawrence Hall of Science #5200
Berkeley, CA 94720-5200
www.lawrencehallofscience.org/planetarium

Glossary

Asteroid belt The region between Mars and Jupiter, about 2.7 AU from the Sun, where most of the asteroids that orbit the Sun can be found. This region is considered the Main Asteroid Belt; other asteroids, called the Trojan Asteroids, are located in Jupiter's orbit.

Asteroids Asteroids are rocky bodies, mostly located between Mars and Jupiter in the Asteroid belt, left over from the formation of the solar system about 4.6 billion years ago.

Astronomical unit (AU) The average distance between the Sun and Earth. One AU is equivalent to 92,955,800 miles (149,597,870 km).

Atmosphere The air surrounding Earth; the gas surrounding a planet.

Aurora australis The light display seen over the South Pole created by charged particles in solar winds that are channeled through Earth's magnetic field. NASA's orbiting Polar Spacecraft recently discovered that the aurora australis and the aurora borealis are nearly mirror images of each other.

Aurora borealis The brilliant light display seen over the North Pole created by charged particles in solar winds that are channeled through Earth's magnetic field. NASA's orbiting Polar Spacecraft recently discovered that the aurora australis and the aurora borealis are nearly mirror images of each other.

Binary Coined by William Herschel, often refers to a binary star system, meaning a pair of stars where mutual gravity causes the stars to follow an elliptical orbit around a "balance point" or a shared center of gravity, as if they were tied together.

Black hole A region of space with an extremely high concentration of mass and such a strong gravitational pull that even light cannot escape. Black holes are thought to form when very massive stars collapse.

Comet Dirty ice left over from the formation of the solar system 4.6 billion years ago. Short-period comets have orbits that are easy to predict, while long-period comets often come from the region past Neptune and might take as long as 30 million years to complete one trip around the Sun.

Corona The Sun's outermost atmosphere. The corona is the light that is visible during a total eclipse when the Moon covers the body of the Sun.

Coronal mass ejections Bubble-shaped bursts of solar winds coming from the solar corona, they are often associated with large flares. These bursts can often cause huge disturbances in Earth's magnetosphere.

Dark matter Made up of particles that do not absorb, reflect, or emit light, it cannot be detected by observing electromagnetic radiation. Dark matter is material that cannot be seen directly; its existence is known because of the effect it has on objects that we can observe directly.

Emission nebulae Hot young stars heat up the gas in the surrounding area, which causes electrons to bump into one another and omits light. The location of the electrons in the orbits dictates the wavelength of the light it omits. Scientists study these wavelengths to determine the chemical content of the nebula. An Emission nebula is about 90 percent hydrogen.

Exoplanets Also known as "extrasolar planets," these are planets orbiting a star other than the Sun.

Exterior planets The planets that orbit past the main asteroid belt, which include: Jupiter, Saturn, Uranus, and Neptune (Pluto, which is now classed as a dwarf planet, is no longer included).

General theory of relativity The theories formed by Albert Einstein that redefined the fundamental concepts of gravity, energy, space, time, and matter.

Globular clusters A group of densely packed stars in a relatively spherical shape.

Heliocentric Relating to the center of the Sun; measured from or referring to the center of the Sun.

Heliosphere The area of space that is directly influenced by the Sun and its powerful winds.

Light pollution Artificial light, such as streetlights, that obscures the night sky.

Messier number Charles Messier, a French astronomer born in 1730, searched for comets. He assigned numbers to objects in the night sky that he identified as not being comets. Many sky objects are still referred to by their original Messier numbers.

Meteorite Meteors that do not burn up in Earth's atmosphere and land on Earth's surface.

Meteors Called "shooting stars," meteors are bits of material, often left behind by a comet, that ignite and burn up as they pass into Earth's atmosphere.

Milky Way The large spiral galaxy that contains over 200 billion stars and is home to the Sun, located in the Orion arm.

New moon Part of the cycle of the Moon's phases, which goes from New Moon to Full Moon back to New again over the course of 29.5 days. A New Moon appears as a sliver in the sky.

Oort cloud An enormous spherical cloud surrounding our solar system that contains billions of icy objects and is thought to be the region where comets come from.

Open clusters Groups of stars that are physically related and held together by gravitational forces.

Parallax The method called stellar parallax is used to estimate the distance to the nearest stars, and depends solely on Earth's orbit around the Sun in order to use geometry to calculate the distance. This method does not work for stars located at a distance greater than 100 light-years.

Parsec The unit used to describe the distance to stars and galaxies. One parsec equals 3.26 light-years and distances are often expressed in kiloparsecs (1 kiloparsec = 1,000 parsecs).

Solar flares Intense and sudden outbursts of the lower layers of the Sun's atmosphere, usually near a sunspot group.

Solar wind Particles, mostly composed of protons and electrons, that are accelerated to speeds that allow them to escape the Sun's gravitational field by the intense heat of the Sun's corona.

Spiral galaxy Galaxies composed of arms full of stars stem and twist around a bright, central nucleus, creating the characteristic pinwheel shape.

Van Allen belts Regions of high-energy particles, mostly protons and electrons, held in doughnut-shaped regions in the high altitudes of Earth's magnetic field.

Variable star Stars whose observed light changes in intensity. The changes can be irregular, regular, or semiregular.

Index

285

Credits

Photo Credits

Backyard Guide to the Night Sky

About the Authors

Howard Schneider is a veteran reporter who is a regular contributor on science and health for the *Washington Post.*

Sandy Wood is the host of "Star-Date," the radio show produced by the University of Texas McDonald Observatory. Ms. Wood's show is aimed at amateur astronomers and anyone interested in the night sky.

Patricia Daniels has written extensively on history and science, including the *National Geographic Encyclopedia of Space, Body: The Complete Human,* and *My First Pocket Guide: Constellations.*

John Scalzi is an award-winning science fiction author and writer of many nonfiction works, including *The Rough Guide to the Universe.*

Alan Dyer is one of Canada's best-known astronomers. He is author of several astronomy books and serves as a contributing editor to *Sky & Telescope* magazine.

Backyard Guide to the Night Sky

Foreword by Sandy Wood
By Howard Schneider; Essays by Patricia Daniels,
John Scalzi, and Alan Dyer

Published by the National Geographic Society
John M. Fahey, Jr., *President and Chief Executive Officer*
Gilbert M. Grosvenor, *Chairman of the Board*
Tim T. Kelly, *President, Global Media Group*
John Q. Griffin, *President, Publishing*
Nina D. Hoffman, *Executive Vice President;*
      President, Book Publishing Group

Prepared by the Book Division
Kevin Mulroy, *Senior Vice President and Publisher*
Leah Bendavid-Val, *Director of Photography Publishing*
      and Illustrations
Marianne R. Koszorus, *Director of Design*
Barbara Brownell Grogan, *Executive Editor*
Elizabeth Newhouse, *Director of Travel Publishing*
Carl Mehler, *Director of Maps*

Staff for This Book
NATIONAL GEOGRAPHIC SOCIETY
Amy Briggs, *Editor*
Robert Burnham, *Consultant*
Judith Klein, *Copy Editor*
Elizabeth Thompson, *Researcher*
Susan Blair, *Illustrations Editor*
Melissa Farris, *Art Director*
Wil Tirion, *Star Charts*
Michael McNey, *Map Production*
Connie D. Binder, *Indexer*
Mike Horenstein, *Production Project Manager*
Jennifer A. Thornton, *Managing Editor*
R. Gary Colbert, *Production Director*

THE BROWN REFERENCE GROUP LTD
Sophie Mortimer, *Picture Manager*
David Poole, *Design Manager*
Tony Truscott, *Designer*
Andrew Webb, *Picture Researcher*
Lindsey Lowe, *Editorial Director*
Tim Harris, *Managing Editor*

Manufacturing and Quality Management
Christopher A. Liedel, *Chief Financial Officer*
Phillip L. Schlosser, *Vice President*
Chris Brown, *Technical Director*
Nicole Elliott, *Manager*
Monika D. Lynde, *Manager*
Rachel Faulise, *Manager*

Founded in 1888, the National Geographic Society
is one of the largest nonprofit scientific and educa-
tional organizations in the world. It reaches more
than 285 million people worldwide each month
through its official journal, *National Geographic,* and
its four other magazines; the National Geographic
Channel; television documentaries; radio programs;
films; books; videos and DVDs; maps; and interactive
media. National Geographic has funded more than
8,000 scientific research projects and supports an
education program combating geographic illiteracy.

For more information, please call 1-800-NGS LINE
(647-5463) or write to the following address:

National Geographic Society
1145 17th Street N.W.
Washington, D.C. 20036-4688 U.S.A.

Visit us online at www.nationalgeographic.com

For information about special discounts for bulk pur-
chases, please contact National Geographic Books
Special Sales: ngspecsales@ngs.org

For rights or permissions inquiries, please contact
National Geographic Books Subsidiary Rights:
ngbookrights@ngs.org

ISBN: 978-1-4262-0281-0 (trade paper)
ISBN: 978-1-4262-0538-5 (hardcover)
ISBN: 978-1-4262-0539-2 (deluxe)

Printed in China